T0221271

Recent Advancements in Software Reliability Assurance

Advances in Mathematics and Engineering
Series Editor: Mangey Ram

The main aim of this focus book series is to publish the original articles that bring up the latest development and research in mathematics and its applications. The books in this series are in short form, ranging between 20,000 and 50,000 words or 100 to 125 printed pages, and encompass a comprehensive range of mathematics and engineering areas. It will include, but won't be limited to, mathematical engineering sciences, engineering and technology, physical sciences, numerical and computational sciences, space sciences, and meteorology. The books within the series line will provide professionals, researchers, educators, and advanced students in the field with an invaluable reference into the latest research and developments.

Recent Advancements in Software Reliability Assurance
Edited by Adarsh Anand and Mangey Ram

Recent Advancements in Software Reliability Assurance

Edited by
Adarsh Anand
Mangey Ram

CRC Press
Taylor & Francis Group
Boca Raton London New York

CRC Press is an imprint of the
Taylor & Francis Group, an **informa** business

CRC Press
Taylor & Francis Group
6000 Broken Sound Parkway NW, Suite 300
Boca Raton, FL 33487-2742

First issued in paperback 2020

© 2019 by Taylor & Francis Group, LLC
CRC Press is an imprint of Taylor & Francis Group, an Informa business

No claim to original U.S. Government works

ISBN-13: 978-1-138-36339-7 (hbk)
ISBN-13: 978-0-367-78806-3 (pbk)

This book contains information obtained from authentic and highly regarded sources. Reasonable efforts have been made to publish reliable data and information, but the author and publisher cannot assume responsibility for the validity of all materials or the consequences of their use. The authors and publishers have attempted to trace the copyright holders of all material reproduced in this publication and apologize to copyright holders if permission to publish in this form has not been obtained. If any copyright material has not been acknowledged, please write and let us know so we may rectify in any future reprint.

Except as permitted under U.S. Copyright Law, no part of this book may be reprinted, reproduced, transmitted, or utilized in any form by any electronic, mechanical, or other means, now known or hereafter invented, including photocopying, microfilming, and recording, or in any information storage or retrieval system, without written permission from the publishers.

For permission to photocopy or use material electronically from this work, please access www.copyright.com (http://www.copyright.com/) or contact the Copyright Clearance Center, Inc. (CCC), 222 Rosewood Drive, Danvers, MA 01923, 978-750-8400. CCC is a not-for-profit organization that provides licenses and registration for a variety of users. For organizations that have been granted a photocopy license by the CCC, a separate system of payment has been arranged.

Trademark Notice: Product or corporate names may be trademarks or registered trademarks, and are used only for identification and explanation without intent to infringe.

Library of Congress Cataloging-in-Publication Data

Names: Anand, Adarsh, author. | Ram, Mangey, author.
Title: Recent advancements in software reliability assurance / Adarsh Anand and Mangey Ram.
Description: Boca Raton : Taylor & Francis, 2019. | Includes bibliographical references and index.
Identifiers: LCCN 2019001290| ISBN 9781138363397 (hardback : alk. paper) | ISBN 9780429431630 (ebook)
Subjects: LCSH: Computer software--Reliability. | Software maintenance.
Classification: LCC QA76.76.R44 A53 2019 | DDC 005.3028/7--dc23
LC record available at https://lccn.loc.gov/2019001290

Visit the Taylor & Francis Web site at
http://www.taylorandfrancis.com

and the CRC Press Web site at
http://www.crcpress.com

Contents

Preface, vii

Editors, ix

Contributors, xi

v

Preface

THIS EDITED VOLUME IN the Focus Series, Advances in Mathematics and Engineering, titled *Recent Advancements in Software Reliability Assurance*, includes invited papers that deal with the modeling and analysis of software systems that provide assurance that software has been designed, tested, and checked for security aspects. Such predictions provide a quantitative basis for achieving reliability, risk, and cost goals. The topics covered are organized as follows:

Chapter 1 discusses the determination of complexity of code changes due to the addition of new features, feature improvement, and bug fixing.

Chapter 2 is about stochastic effort estimation modeling that considers quality, cost, and delivery based on maintenance efforts for open source software project management.

Chapter 3 presents a software reliability model for a dynamic development environment.

Vulnerabilities in a software are a major risk to any software firm and if exploited they may cause a massive breach of data and information. Therefore, in Chapter 4, a mathematical framework to model the fixation of vulnerabilities, which are interdependent in nature, is presented.

Change-point is a well-established concept in software reliability studies. Chapter 5 discusses change-point–based software reliability modeling and its application for software development management.

The software upgrading process tends to add more features at frequent time points that lead to an increase in the fault content in the software. Hence, software developers have to continuously check these faults along with the leftover faults from the previous releases. Chapter 6 provides an alternative approach to understand this methodical framework.

Chapter 7 discusses assessing software reliability enhancement achievable through testing.

The editors are grateful to various authors who contributed to this editorial work and the reviewers who helped to improve the quality of the chapters through their comments and suggestions. They deserve significant praise for their assistance. Gratitude also goes out to Navneet Bhatt, research scholar in the Department of Operational Research, who helped in carrying out the entire process in a streamlined manner.

Finally and most importantly, Dr. Adarsh Anand would like to dedicate this editorial book to his parents (Raj Kumari Anand and O. P. Anand), wife (Dr. Deepti Aggrawal), and daughter (Ahana Anand).

Prof. Mangey Ram would like to dedicate this edited book to his family members and friends.

Adarsh Anand
University of Delhi

Mangey Ram
Graphic Era

Editors

Adarsh Anand received his PhD in the area of software reliability assessment and innovation diffusion modeling in marketing. Presently he is working as an assistant professor in the Department of Operational Research, University of Delhi, India. He was as a Young Promising Researcher in the field of Technology Management and Software Reliability by the Society for Reliability Engineering, Quality and Operations Management (SREQOM) in 2012. He is a lifetime member of SREQOM. He is also on the editorial board of the *International Journal of System Assurance and Engineering Management* (Springer) and the *International Journal of Mathematical, Engineering and Management Sciences.* He is one of the editors of the book *System Reliability Management: Solutions & Technologies* (CRC Press/Taylor & Francis Group). He has guest-edited several special issues for journals of international repute. He has published articles in journals of national and international repute. His research interests include software reliability growth modeling, modeling innovation adoption and successive generations in marketing, and social network analysis.

Mangey Ram received a PhD in mathematics with a minor in computer science from G. B. Pant University of Agriculture and Technology, Pantnagar, India. He has been a faculty member for about 10 years and has taught several core courses in pure and applied mathematics at the undergraduate, postgraduate, and doctorate levels. He is currently a professor at Graphic Era

(Deemed to be University), Dehradun, India. Before joining Graphic Era, he was a deputy manager (probationary officer) with Syndicate Bank for a short period. He is editor-in-chief of *International Journal of Mathematical, Engineering and Management Sciences*, and a guest editor and member of the editorial board of various journals. He is a regular reviewer for international journals, including those published by IEEE, Elsevier, Springer, Emerald, John Wiley, and Taylor & Francis. He has published 144 research articles in journals published by IEEE, Taylor & Francis, Springer, Elsevier, Emerald, World Scientific, and many other national and international journals of repute, and has also presented his work at national and international conferences. His fields of research are reliability theory and applied mathematics. Ram is a senior member of the IEEE. He is also a life member of the Operational Research Society of India; Society for Reliability Engineering, Quality and Operations Management in India; and the Indian Society of Industrial and Applied Mathematics. He is a member of the International Association of Engineers in Hong Kong, and Emerald Literati Network in the United Kingdom. He has been a member of the organizing committee of a number of international and national conferences, seminars, and workshops. He was conferred with the Young Scientist Award by the Uttarakhand State Council for Science and Technology, Dehradun, in 2009. He was awarded the Best Faculty Award in 2011; Research Excellence Award in 2015; and recently the Outstanding Researcher Award in 2018 for his significant contribution in academics and research at Graphic Era.

Contributors

D. Aggrawal
USME, Delhi Technological
University

Adarsh Anand
University of Delhi

N. R. Barraza
Universidad Nacional de Tres
de Febrero

S. Bharmoria
University of Delhi

S. Das
University of Delhi

S. Inoue
Kansai University

J. Kaur
University of Delhi

Y. K. Malaiya
Computer Science Department
Colorado State University

Y. Minamino
Tottori University

Mangey Ram
Graphic Era University

O. Singh
University of Delhi

H. Sone
Tokyo City University

Y. Tamura
Tokyo City University

S. Yamada
Tottori University

CHAPTER 1

Characterizing the Complexity of Code Changes in Open Source Software

Adarsh Anand, S. Bharmoria, and Mangey Ram

CONTENTS

1.1 INTRODUCTION

In the present-day scenario where consumers' needs are regularly altering, software firms are constantly under pressure to deliver a product that meets all consumer needs and is efficient as well as more reliable than any other similar product on the market with a constraint of time. The fulfillment of these demands received from users requires plenty of source code modifications in the software. These modifications can be due to feature enhancements, inclusion of new features or bug fixes. In open source software (OSS), users are more involved in the process of feature enhancements, inclusion of new features and bug fixes due to the easy access of the source code. The participation of different users in building OSS leads to rapid improvements. The bug fixes, enhancements or inclusion of new features lead to changes in source code, and these vast alterations add to the complexity of code changes. It is the major aspect that decides the quality and reliability of the software. Due to the different code changes contributed by different users/ programmers in the OSS, it becomes a problem to keep footprints of all these changes influencing to complexity. The maintenance job turns out to be quite challenging if these changes are not being accurately noted. The use of a software configuration management repository to record these changes makes maintenance simple. As a result of the presence of a large number of software repositories, a leap in the quantification of software engineering development can be seen. The repositories consist of source code, bugs, interaction between programmers/users, modifications in source code, and so on, and these repositories are used for extracting valuable data for developing the attributes of software. Bugs are reported by bug-reporting systems. Source codes are organized and controlled by source code control systems comprehensively. These repositories are highly informative and are regularly updated by the programmers/users, and the data accounted for by these repositories helps in the study of the source code change process. The study of arrangements of code alteration is the code change process. Source code changes are implemented by the programmers for inclusion of new features, feature enhancements and bug fixing. The repetitive modifications, because of the inclusion of new features, feature enhancement and bug fixing, add to code complexity as a result of the increase in file alterations. Bugs are introduced because of misinteraction or zero interaction between programmers and customers, in addition to software complexity, existence of programming faults, continuously altering demands of customers, timely release constraints, inclusion of new features, feature enhancements, faults under consideration and repairs. The complexity of code changes has been quantified and designed using information theory–based procedures called entropy. In literature, researchers have worked on entropy and it is widely being exercised in software reliability. Shannon (1948, 1951) provided the foundation for the area of information theory and provided the measure for the efficiency of communication channel and entropy. The complexity of code changes matrix is also used for the maintenance of software (Kafura and Reddy, 1987). Entropy is also applied in software reliability to measure the uncertainty, and conditional entropy is applied to establish the uncertainty of different factors (Kamavaram and Goseva-Popstojanova, 2002). Complexity code matrices were considered by Hassan (2009) based on the code change process, and the faults were predicted utilizing the complexity of code changes. A model was suggested by Singh and Chaturvedi (2012) on the basis of entropy for bug prediction utilizing support vector regression. The release time of software has been predicted by Chaturvedi et al. (2013) using the complexity of code changes. It decreases when bugs are corrected, that is, correlation among decline in complexity and bug corrections exist. Then we can say that the decay rate of the complexity of code changes can be studied as the bug discovery/deletion rate. The contribution of modifications caused by feature enhancements, inclusion of new features and bug fixes may obey different decay curves. Different approaches were suggested in the past that quantify the code change metric and decay functions that portray the variability in the decay curves representing the complexity of code changes by Singh and Chaturvedi (2013).

Along with several other measures of entropy, Arora et al. (2014) consider bug prediction models using these measures. Chaturvedi et al. (2014) used measures of entropy in prediction of the complexity of code changes. A defect prediction model was studied to predict how many defects would arise throughout the improvement of software product lines on the basis of entropy (Jeon et al., 2014).

Singh et al. (2015) proposed three models—software reliability growth models, models based on the potential complexity of code changes and models based on the complexity of code changes—to predict the potential bugs in software.

In chapter, we characterize the complexity of code changes into two factors: (1) feature enhancements or inclusion of new features, and (2) bug repairs for an open source software. The model has been verified for three projects of Apache OSS. The implementation of the suggested model has been justified by goodness-of-fit measures.

The article is organized as follows: Section 1.2 provides the suggested model building for characterization of the complexity of code changes. Section 1.3 contributes the data sets, model authentication, and tables and figures backing the suggested model. The chapter is concluded in Section 1.4 followed by references.

1.2 MODEL BUILDING

In this segment, we make use of the model proposed by Chaturvedi et al. (2014). It is very interesting to note that this model is analogous to the software reliability model used by Anand et al. (2013). We have taken b_1 as the diffusion rate of entropy/uncertainty/complexity of code changes caused by inclusion of new feature or feature enhancements, and b_2 as the diffusion rate of entropy/uncertainty/complexity of code changes caused by bug repairs. The proposed model is given in Equation 1.1 on the basis of corresponding assumptions:

1. Potential entropy/the complexity of code changes ($\overline{H}$) is constant.

2. The diffusion of entropy caused by inclusion of new features/feature enhancements is independent.

3. At $t = 0$, the file has zero change and entropy/complexity of code changes is zero.

The complexity of code change diffusion per unit time can be written as follows:

$$\frac{d(H(t))}{dt} = b_1(\overline{H} - H(t)) + b_2 \frac{H(t)}{\overline{H}}(\overline{H} - H(t)) \tag{1.1}$$

where $\overline{H}$ is the potential entropy/complexity of code changes to be diffused in software over a period of time and $H(t)$ is the amount of entropy/complexity of code changes at any given time t. Solving the preceding differential equation with initial conditions at $t = 0$ and $H(0) = 0$, we get

$$H(t) = \overline{H} \left[\frac{1 - e^{-(b_1 + b_2)t}}{1 + \frac{b_2}{b_1} e^{-(b_1 + b_2)t}} \right] \tag{1.2}$$

Equation 1.2 provides the complexity of code changes at any given time t. Further, we can take $\phi = b_1 + b_2$ as the complexity of code changes diffusion rate because of the changes arise in source code and $\beta = (b_2/b_1)$ as constant.

In order to make use of the model, we would first study this model based on its cumulative distribution function, that is, Equation 1.1 can be rewritten as

$$\frac{dF(t)}{dt} = [b_1 + b_2 F(t)][1 - F(t)] \tag{1.3}$$

where $F(t) = (H(t)/\overline{H})$ is cumulative complexity of code changes until time t and $f(t) = ((dF(t))/dt)$ is the rate of complexity of code changes at time t. If $\overline{H}$ is the potential complexity of code changes to be diffused in a software over a period of time, then the cumulative amount of complexity of code changes at time t is $\overline{H}F(t)$. Integration of Equation 1.3 results in the S-shaped cumulative complexity of code changes distribution, $F(t)$. Further the differentiation of $F(t)$ gives noncumulative complexity of code changes distribution. The distribution is thus given by

$$F(t) = \left(\frac{1 - e^{-(b_1 + b_2)t}}{1 + \dfrac{b_2}{b_1} e^{-(b_1 + b_2)t}} \right) \tag{1.4}$$

and

$$f(t) = \frac{b_1 (b_1 + b_2)^2 e^{-(b_1 + b_2)t}}{(b_1 + b_2 e^{-(b_1 + b_2)t})^2} \tag{1.5}$$

$f(t)$ is the noncumulative complexity of code changes distribution. As solved, its peak, $f(T^*)$ or $F(T^*)$, at time T^* occurs when

$$T^* = -\frac{1}{(b_1 + b_2)} \ln\left(\frac{b_1}{b_2}\right), \tag{1.6}$$

$$F(T^*) = \frac{1}{2} - \frac{b_1}{2b_2}, \tag{1.7}$$

and

$$f(T^*) = \frac{1}{4b_2}(b_1 + b_2)^2 \tag{1.8}$$

We note from Equation 1.5 that $f(t = 0) = f(t = 2T^*) = b_1$. The term $b_1[1 - F(b)]$ in Equation 1.3 signifies complexity of code changes caused by feature enhancements or inclusion of new features. On the other hand, the term $b_2 F(t)[1 - F(t)]$ in Equation 1.3 signifies complexity of code changes caused by bug fixing. This is portrayed in Figure 1.1.

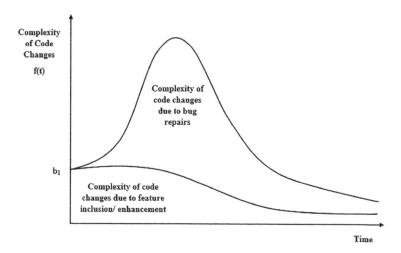

FIGURE 1.1 Complexity of code changes due to feature inclusion or enhancement and bug repairs.

Since our goal is to characterize complexity, a model is suggested to quantify the different complexity of code changes present in software. We have assumed $F(t)$ is the cumulative complexity of code changes until time t. Since, it can be bifurcated into two factors, that is, inclusion of new features/ feature enhancements and bug repairs, $F(t)$ can be assumed to contain two factors: $F_1(t)$ corresponding to complexity caused by feature enhancements or inclusion of new features, and $F_2(t)$ corresponding to complexity caused by bug fixing, where the total fraction of complexity caused by feature enhancements or inclusion of new features $F_1(t)$ between any two time periods, say t_0 and $t_F(t_F > t_0)$, is given by

$$F_1(t) = b_1 \int_{t_0}^{t_F} [1 - F(t)]dt \qquad (1.9)$$

where $F(t)$ is given by Equation 1.4. Hence, $F_1(t)$ can be deduced as

$$
\begin{aligned}
F_1(t) &= b_1 \int_{t_0}^{t_F} \left[1 - \frac{1 - e^{-(b_1 + b_2)t}}{1 + \left(\dfrac{b_2}{b_1}\right) e^{-(b_1 + b_2)t}} \right] dt \\
&= b_1 \int_{t_0}^{t_F} \left[\frac{b_1 + b_2 e^{-(b_1 + b_2)t} - b_1 \left(1 - e^{-(b_1 + b_2)t}\right)}{b_1 + b_2 e^{-(b_1 + b_2)t}} \right] dt \quad (1.10) \\
&= b_1 \int_{t_0}^{t_F} \left[\frac{(b_1 + b_2) e^{-(b_1 + b_2)t}}{b_1 + b_2 e^{-(b_1 + b_2)t}} \right] dt
\end{aligned}
$$

Substituting $Z = e^{-(b_1 + b_2)t}$ or $dZ = -(b_1 + b_2)e^{-(b_1 + b_2)t}\, dt$ in the preceding equation:

$$
\begin{aligned}
F_1(t) &= -b_1 \int_{z_0}^{z_f} \frac{dZ}{b_1 + b_2 Z} \\
&= \frac{-b_1}{b_2} \ln[(b_1 + b_2 Z)]_{z_0}^{z_f} \quad (1.11) \\
&= \frac{b_1}{b_2} \ln \left[\frac{b_1 + b_2 e^{-(b_1 + b_2)t_0}}{b_1 + b_2 e^{-(b_1 + b_2)t_f}} \right]
\end{aligned}
$$

Substituting $t_0 = 0$, $t_f = t$ in the preceding equation, we get

$$
F_1(t) = \frac{b_1}{b_2} \ln \left[\frac{1 + \dfrac{b_2}{b_1}}{1 + \dfrac{b_2}{b_1} e^{-(b_1 + b_2)t}} \right] \quad (1.12)
$$

Equation 1.12 gives us the total fraction of complexity caused by feature enhancements or inclusion of new features, and to convert the corresponding fraction into the amount of complexity caused by feature enhancements or inclusion of new features, we multiply by $\overline{H}$. Similarly, $\overline{H}$ is multiplied by $F_2(t)$, that is, the fraction of complexity caused by bug fixing to convert it into amount of complexity caused by bug fixing, and it is derived by subtracting $F_1(t)$ from 1 as follows:

$$F_2(t) = 1 - F_1(t) \tag{1.13}$$

$$F_2(t) = \left[1 - \frac{b_1}{b_2} \ln \left[\frac{1 + \frac{b_2}{b_1}}{1 + \frac{b_2}{b_1} e^{-(b_1 + b_2)t}} \right] \right] \tag{1.14}$$

1.3 DATA SET AND MODEL VALIDATION

In order to examine our suggested model, we have used the data gathered from three projects, namely Avro (DS-I), Hive (DS-II) and Pig (DS-III) of Apache OSS (Singh and Sharma, 2014). We have used each data set to estimate the complexity. Table 1.1 refers to the estimated parameters of the model for each data set. Tables 1.2 through 1.4 are composed of actual and predicted values

TABLE 1.1 Parameter Estimates for Three Data Sets

Project	Parameter	Estimate
DS-I	$\overline{H}$	94.044
	b_1	0.029
	b_2	0.111
DS-II	$\overline{H}$	215.024
	b_1	0.018
	b_2	0.023
DS-III	$\overline{H}$	101.267
	b_1	0.030
	b_2	0.089

TABLE 1.2   Characterization of Complexity of Code Changes for DS-I

| Time | Actual Entropy | Predicted Entropy | Entropy Due to | |
			Feature Inclusion or Enhancement	Bug Repairs
1	0.989	2.799	2.687	0.112
2	4.866	5.818	5.288	0.531
3	8.722	9.052	7.797	1.255
4	12.545	12.490	10.208	2.281
5	16.404	16.114	12.516	3.598
6	20.343	19.902	14.714	5.187
7	24.240	23.827	16.800	7.026
8	28.180	27.855	18.770	9.086
9	32.122	31.952	20.620	11.332
10	36.085	36.077	22.351	13.727
11	40.011	40.191	23.961	16.230
12	43.960	44.254	25.452	18.802
13	47.884	48.229	26.826	21.403
14	51.799	52.080	28.086	23.994
15	55.778	55.780	29.237	26.542
16	59.684	59.301	30.284	29.017
17	62.684	62.626	31.231	31.395

of entropy along with its characterization into inclusion of new features or feature enhancements and bug repairs for each data set.

Table 1.2 represents the separated values pertaining to the two classes under consideration (new features/feature enhancements and bug repairs). It can be observed from the table that complexity caused by feature enhancements or inclusion of new features in starting time points is more than the complexity caused by bug fixing, but as time progresses the bug repairs contribute to the complexity to a very good extent. If we take $t = 3$, the complexity caused by feature enhancements/inclusion of new features and bug repairs is 7.797 and 1.255 respectively, and for $t = 16$ the complexity caused by feature enhancements/inclusion of new features and bug repairs is 30.284 and 29.017 respectively.

TABLE 1.3   Characterization of Complexity of Code Changes for DS-II

| Time | Actual Entropy | Predicted Entropy | Entropy Due to | |
			Feature Inclusion or Enhancement	Bug Repairs
1	3.919	3.842	3.836	0.007
2	7.795	7.700	7.601	0.099
3	11.645	11.571	11.296	0.275
4	15.499	15.452	14.921	0.530
5	19.391	19.339	18.476	0.863
6	23.274	23.229	21.959	1.270
7	27.114	27.119	25.372	1.747
8	30.924	31.006	28.714	2.292
9	34.796	34.887	31.986	2.902
10	38.682	38.759	35.187	3.572
11	42.526	42.617	38.318	4.300
12	46.400	46.460	41.379	5.082
13	50.215	50.284	44.370	5.914
14	54.112	54.087	47.292	6.795
15	57.990	57.865	50.145	7.719
16	61.833	61.615	52.930	8.685
17	65.677	65.335	55.647	9.688
18	68.587	69.022	58.297	10.725

In Table 1.3, if we take $t = 8$, then we see that actual and predicted entropy are 30.924 and 31.006 respectively, and the complexity caused by feature enhancements/inclusion of new features is 28.714 and caused by bug repairs is 2.292. From Table 1.3, we are able to conclude that contribution of inclusion of new features/feature enhancements is more complex than bug repairs.

In Table 1.4, if we take $t = 13$, then we see that actual and predicted entropy are 48.939 and 49.110 respectively, and complexity caused by inclusion of new feature or enhancement is 30.325 and due to bug repairs is 18.785. From the Table 1.4, we are able to conclude that contribution of feature enhancements/ inclusion of new features is more complex than bug repairs

TABLE 1.4 Characterization of Complexity of Code Changes for DS-III

Time	Actual Entropy	Predicted Entropy	Entropy Due to Feature Inclusion or Enhancement	Bug Repairs
1	1.982	3.150	2.992	0.158
2	5.919	6.470	5.887	0.583
3	9.807	9.950	8.680	1.270
4	13.725	13.580	11.368	2.212
5	17.627	17.340	13.946	3.394
6	21.513	21.210	16.410	4.800
7	25.390	25.160	18.757	6.403
8	29.295	29.170	20.985	8.185
9	33.236	33.210	23.093	10.117
10	37.163	37.260	25.080	12.180
11	41.085	41.270	26.947	14.323
12	45.026	45.230	28.695	16.535
13	48.939	49.110	30.325	18.785
14	52.876	52.890	31.841	21.049
15	56.799	56.530	33.245	23.285

TABLE 1.5 Goodness-of-Fit Measures of Three Data Sets

Project	R^2	MSE	Bias	Variation	RMSPE
DS-I	0.999	0.315	−0.121	0.565	0.577
DS-II	1	0.024	0.010	0.160	0.160
DS-III	1	0.143	−0.077	0.383	0.390

in starting time points, but as time progresses, the bug repairs contribution increases in complexity to an extent.

Table 1.5 refers to the goodness-of-fit measures of the model used for each data set. An R^2 value closer to 1 for each data set certifies that our quantified model is a good fit.

1.4 CONCLUSION

There has been a rapid increase in the use of OSS. The operating users/programmers are adding to the evolvement of software by bug repairs, inclusion of new features and feature enhancements.

In this chapter, we have recommended a quantification technique through an established model and characterized the complexity in OSS. We categorized the complexity into two factors, that is, feature enhancements/inclusion of new features and bug repairs. We have validated the suggested model using three projects of Apache OSS. Implementation of the suggested model has been estimated on the basis of several goodness-of-fit criteria and they have been found to fit the real data. The suggested model to characterize complexity of code changes of open source software will benefit in development of software.

REFERENCES

Anand, A., D. Aggrawal, S. Das, and D. Dhiman. "Computation of discrepant faults using flexible software reliability growth modelling framework." *Communications in Dependability and Quality Management (CDQM): An International Journal*, 16(2), 15–27, 2013.

Arora, H. D., V. Kumar, and R. Sahni. "Study of bug prediction modeling using various entropy measures: A theoretical approach." In: *2014 3rd International Conference on Reliability, Infocom Technologies and Optimization (ICRITO) (Trends and Future Directions)*, pp. 1–5. IEEE, 2014.

Chaturvedi, K. K., P. Bedi, S. Misra, and V. B. Singh. "An empirical validation of the complexity of code changes and bugs in predicting the release time of open source software." In: *2013 IEEE 16th International Conference on Computational Science and Engineering (CSE)*, pp. 1201–1206. IEEE, 2013.

Chaturvedi, K. K., P. K. Kapur, S. Anand, and V. B. Singh. "Predicting the complexity of code changes using entropy based measures." *International Journal of System Assurance Engineering and Management*, 5(2), 155–164, 2014.

Hassan, A. E. "Predicting faults using the complexity of code changes." In: *Proceedings of the 31st International Conference on Software Engineering*, pp. 78–88. IEEE Computer Society, 2009.

Jeon, C. K., C. Byun, N. H. Kim, and H. P. In. "An entropy based method for defect prediction in software product lines." *International Journal of Multimedia and Ubiquitous Engineering*, 9(3), 375–377, 2014.

Kafura, D., and G. R. Reddy. "The use of software complexity metrics in software maintenance." *IEEE Transactions on Software Engineering*, 3, 335–343, 1987.

Kamavaram, S., and K. Goseva-Popstojanova. "Entropy as a measure of uncertainty in software reliability." In: *13th International Symposium Software Reliability Engineering*, pp. 209–210. 2002.

Shannon, C. E. "A mathematical theory of communication." *Bell System Technical Journal*, 27(3), 379–423, 1948.

Shannon, C. E. "Prediction and entropy of printed English." *Bell System Technical Journal*, 30(1), 50–64, 1951.

Singh, V. B., and K. K. Chaturvedi. "Entropy based bug prediction using support vector regression." In: *2012 12th International Conference on Intelligent Systems Design and Applications (ISDA)*, pp. 746–751. IEEE, 2012.

Singh, V. B., and K. K. Chaturvedi. "Improving the quality of software by quantifying the code change metric and predicting the bugs." In: *International Conference on Computational Science and Its Applications*, pp. 408–426. Springer, Berlin, Heidelberg, 2013.

Singh, V. B., K. K. Chaturvedi, S. K. Khatri, and V. Kumar. "Bug prediction modeling using complexity of code changes." *International Journal of System Assurance Engineering and Management*, 6(1), 44–60, 2015.

Singh, V. B., and M. Sharma. "Prediction of the complexity of code changes based on number of open bugs, new feature and feature improvement." In: *2014 IEEE International Symposium on Software Reliability Engineering Workshops (ISSREW)*, pp. 478–483. IEEE, 2014.

CHAPTER 2

Stochastic Effort Estimation for Open Source Projects

Y. Tamura, H. Sone, and S. Yamada

CONTENTS

2.1 INTRODUCTION

The method of earned value management (EVM) (Fleming and Koppelman, 2010) is applied to the actual software projects under various IT companies. Also, much open source software (OSS) is developed and managed by using fault big data recorded on bug tracking systems. At present, OSS is used in various application areas, because OSS is useful for many users to make cost reductions, standardization, and quick delivery. The methods for reliability assessment of OSS have been proposed by several researchers (Norris 2004, Yamada and Tamura 2016, Zhou and Davis 2005). Also, our research group has proposed and discussed the method of reliability assessment for proprietary software and OSS (Kapur et al. 2011, Lyu 1996, Musa et al. 1987, Yamada 2014). However, research focused on the software effort expenditures of OSS has not been proposed. In particular, it is important to appropriately control the quality according to the progress status of the OSS project. Also, the appropriate control of management efforts for OSS will indirectly link to the quality, reliability, and cost, because much OSS is developed and maintained by several developers with many OSS users considering the quality of the OSS project.

In this chapter, we remake the OSS effort estimation model by using the conventional stochastic differential equation model. Also, this chapter proposes a useful method of OSS project management based on earned value analysis considering the irregular fluctuation of performance resulting from the characteristics of OSS development and management. Moreover, we devise the noisy terms included in the proposed effort estimation model. Furthermore, several numerical examples of earned value analysis and noisy characteristics based on the proposed method are shown by using the effort data under an actual OSS project.

2.2 STOCHASTIC WIENER PROCESS MODELS FOR OSS EFFORT ESTIMATION

Considering the characteristic of the operation phase of OSS projects, the time-dependent expenditure phenomena of maintenance efforts keep an irregular state in the operation phase, because there is variability among the levels of developer skill and development environment, and the OSS is developed and maintained by several developers and users. Then, the time-dependent expenditure phenomena of the maintenance effort become unstable.

The operation phases of many OSS projects are influenced from external factors by triggers such as the difference of skill, time lag of development, and maintenance. Considering these points, we apply the stochastic differential equation model to manage the OSS project. Then, let $\Lambda(t)$ be the cumulative maintenance effort expenditures up to operational time t ($t \geq 0$) in the OSS development project. Suppose that $\Lambda(t)$ takes on continuous real values. Since the estimated maintenance efforts are observed during the operational phase of the OSS project, $\Lambda(t)$ gradually increases as the operational procedures go on. Based on the software reliability growth modeling approach (Kapur et al. 2011, Lyu 1996, Musa et al. 1987, Yamada 2014), the following linear differential equation in terms of maintenance effort can be formulated:

$$\frac{d\Lambda(t)}{dt} = \beta(t)\{\alpha - \Lambda(t)\}, \qquad (2.1)$$

where $\beta(t)$ is the increase rate of maintenance effort at operational time t and a non-negative function, and α means the estimated maintenance effort required until the end of operation.

Therefore, we extend Equation 2.1 to the following stochastic differential equation with Brownian motion (Arnold 1974, Wong 1971):

$$\frac{d\Lambda(t)}{dt} = \{\beta(t) + \sigma\nu(t)\}\{\alpha - \Lambda(t)\}, \qquad (2.2)$$

where σ is a positive constant representing a magnitude of the irregular fluctuation, and $\nu(t)$ a standardized Gaussian white noise. By using Itô's formula, we can obtain the solution of Equation 2.2 under the initial condition $\Lambda(0) = 0$ as follows:

$$\Lambda(t) = \alpha\left[1 - \exp\left\{-\int_0^t \beta(s)ds - \sigma\omega(t)\right\}\right], \qquad (2.3)$$

where $\omega(t)$ is the one-dimensional Wiener process, which is formally defined as an integration of the white noise $\nu(t)$ with respect to time t. Moreover, we define the increase rate of maintenance effort in case of $\beta(t)$ defined as (Yamada et al. 1994):

$$\int_0^t \beta(s)\,ds \doteq \frac{\dfrac{dF_*(t)}{dt}}{\alpha - F_*(t)}. \tag{2.4}$$

In this chapter, we assume the following equations based on software reliability models $F_*(t)$ as the growth functions of the proposed model. Then, $\Lambda_e(t)$ means the cumulative maintenance effort for the exponential software reliability growth model $F_e(t)$ for $F_*(t)$ in Equation 2.4. Similarly, $\Lambda_s(t)$ is the cumulative maintenance effort for the delayed S-shaped software reliability growth model $F_s(t)$ for $F_*(t)$ in Equation 2.4.

Therefore, the cumulative maintenance effort up to time t is obtained as follows:

$$\Lambda_e(t) = \alpha[1 - \exp\{-\beta t - \sigma\omega(t)\}], \tag{2.5}$$

$$\Lambda_s(t) = \alpha[1 - (1 + \beta t)\exp\{-\beta t - \sigma\omega(t)\}]. \tag{2.6}$$

Moreover, the estimated maintenance effort required until the end of operation is obtained as follows:

$$\Lambda_{re}(t) = \alpha\exp\{-\beta t - \sigma\omega(t)\}, \tag{2.7}$$

$$\Lambda_{rs}(t) = \alpha(1 + \beta t)\exp\{-\beta t - \sigma\omega(t)\}. \tag{2.8}$$

In this model, we assume that the parameter σ depends on several noises by external factors from several triggers in OSS projects. Then, from Equations 2.5 and 2.6, the expected cumulative maintenance effort expenditures spent up to time t are respectively obtained as follows (Tamura et al. 2018):

$$E[\Lambda_e(t)] = \alpha\left[1 - \exp\left\{-\beta t + \frac{\sigma^2}{2}t\right\}\right], \tag{2.9}$$

$$E[\Lambda_s(t)] = \alpha \left[1 - (1 + \beta t)\exp\left\{ -\beta t + \frac{\sigma^2}{2} t \right\} \right]. \qquad (2.10)$$

In the case of $\Lambda_e(t)$, log$\{\alpha - \Lambda_e(t)\}$ defined in Equation 2.5 takes the Gaussian process because the Wiener process $w(t)$ is the Gaussian process. Then, the expected value and variance of log$\{\alpha - \Lambda_e(t)\}$ are respectively given as

$$E[\log\{\alpha - \Lambda_e(t)\}] = \log\alpha - \beta t, \qquad (2.11)$$

$$\mathrm{Var}[\log\{\alpha - \Lambda_e(t)\}] = \sigma^2 t. \qquad (2.12)$$

Also, we can obtain as follows:

$$\Pr[\log\{\alpha - \Lambda_e(t)\} \le x] = \Phi\left(\frac{x - \log\alpha + \beta t}{\sigma\sqrt{t}} \right), \qquad (2.13)$$

where Φ means the following standard normal distribution function:

$$\Phi(x) = \frac{1}{\sqrt{2\pi}} \int_{-\infty}^{x} \exp\left(-\frac{y^2}{2} \right) dy. \qquad (2.14)$$

Therefore, the transition probability $\Lambda_e(t)$ is derived as follows:

$$\Pr[\Lambda_e(t) \le n \mid \Lambda_e(0) = 0] = \Phi\left(\frac{\log\dfrac{\alpha}{\alpha - n} - \beta t}{\sigma\sqrt{t}} \right). \qquad (2.15)$$

Similarly, in case of $\Lambda_s(t)$, log$\{\alpha - \Lambda_s(t)\}$ defined in Equation 2.6 takes the Gaussian process because Wiener process $w(t)$ is the Gaussian process. Then, the expected value and variance of $\log\{\alpha - \Lambda_s(t)\}$ are respectively given as:

$$E[\log\{\alpha - \Lambda_s(t)\}] = \log\alpha - \beta t + \log(1 + \beta t), \quad (2.16)$$

$$\mathrm{Var}[\log\{\alpha - \Lambda_s(t)\}] = \sigma^2 t. \quad (2.17)$$

Also, we can obtain the following:

$$\Pr[\log\{\alpha - \Lambda_s(t)\} \le x] = \Phi\left(\frac{x - \log\alpha + \beta t - \log(1 + \beta t)}{\sigma\sqrt{t}}\right).$$

$$(2.18)$$

Therefore, the transition probability $\Lambda_s(t)$ is derived as follows:

$$\Pr[\Lambda_s(t) \le n \mid \Lambda_s(0) = 0] = \Phi\left(\frac{\log\dfrac{\alpha}{\alpha - n} - \beta t + \log(1 + \beta t)}{\sigma\sqrt{t}}\right).$$

$$(2.19)$$

2.3 STOCHASTIC JUMP MODELS FOR OSS EFFORT ESTIMATION

The jump term can be added to the proposed stochastic differential equation models in order to incorporate the irregular state around the time t by various external factors in the maintenance phase of the OSS project. Then, the jump-diffusion process is given as follows (Merton 1976):

$$d\Lambda_j(t) = \left\{\beta(t) - \frac{1}{2}\sigma^2\right\}\{\alpha - \Lambda(t)\}dt + \sigma\{\alpha - \Lambda_j(t)\}d\omega(t)$$

$$+ d\left\{\sum_{i=1}^{Y_t(\lambda)}(V_i - 1)\right\}, \quad (2.20)$$

where $Y_t(\lambda)$ is a Poisson point process with parameter λ at operation time t. Also, $Y_t(\lambda)$ is the number of occurred jumps and λ the jump rate. $Y_t(\lambda)$, $\omega(t)$, and V_i are assumed to be mutually independent. Moreover, V_i is ith jump range.

By using Itô's formula, the solution of the former equation can be obtained as follows (Yamada and Tamura 2018):

$$\Lambda_{je}(t) = \alpha \left[1 - \exp\left\{-\beta t - \sigma\omega(t) - \sum_{i=1}^{Y_t(\lambda)} \log V_i \right\}\right], \quad (2.21)$$

$$\Lambda_{js}(t) = \alpha \left[1 - (1+\beta t)\exp\left\{-\beta t - \sigma\omega(t) - \sum_{i=1}^{Y_t(\lambda)} \log V_i \right\}\right]. \quad (2.22)$$

Similarly, the estimated maintenance effort required until the end of operation is obtained as follows:

$$\Lambda_{rje}(t) = \alpha\exp\left\{-\beta t - \sigma\omega(t) - \sum_{i=1}^{Y_t(\lambda)} \log V_i \right\}, \quad (2.23)$$

$$\Lambda_{rjs}(t) = \alpha(1+\beta t)\exp\left\{-\beta t - \sigma\omega(t) - \sum_{i=1}^{Y_t(\lambda)} \log V_i \right\}. \quad (2.24)$$

2.4 EFFORT ASSESSMENT MEASURES FOR EVM IN OSS

The method of EVM is often used for the project management under various IT companies. Generally, the EVM is applied to common software development projects (Fleming and Koppelman 2010). However, it is difficult to directly apply EVM to the actual OSS project, because the development cycle of the OSS project is different from the typical software development paradigm. As is characteristic of OSS, the OSS development project is managed by using the actual bug tracking system. This chapter proposes the method of earned value analysis for OSS projects by using the data sets obtained from the bug tracking system.

Considering the earned value analysis for OSS, we assume the following terms as the EVM for OSS (Tamura et al. 2018):

- *Actual Cost (AC)*: Cumulative maintenance effort up to operational time t considering the reporter and assignee

- *Earned Value (EV)*: Cumulative maintenance effort up to operational time t considering the reporter

- *Cost Variance (CV)*: Fixing effort required for OSS maintenance up to operational time t, $E[\Theta_e(t)]$ and $E[\Theta_s(t)]$

- *Cost Performance Index (CPI)*: $E[CPI_e(t)]$ and $E[CPI_s(t)]$ obtained from AC and EV

- *Estimate at Completion (EAC)*: $E[EAC_e(t)]$ and $E[EAC_s(t)]$ obtained from AC, EV, CPI, and BAC

- *Estimate to Completion (ETC)*: $E[ETC_e(t)]$ and $E[ETC_s(t)]$ obtained from AC, EV, CPI, BAC, and EAC

- *Budget at Completion (BAC)*: Planned Value (PV) in the end point as the specified goal of OSS project

Then, the expected fixing effort required for OSS maintenance up to operational time t in case of $\Lambda_e(t)$ and $\Lambda_s(t)$ can be formulated as

$$E[\Theta_e(t)] = E\left[\Lambda_{re}^r(t)\right] - E\left[\Lambda_{re}^{ra}(t)\right], \qquad (2.25)$$

$$E[\Theta_s(t)] = E\left[\Lambda_{rs}^r(t)\right] - E\left[\Lambda_{rs}^{ra}(t)\right], \qquad (2.26)$$

where $E\left[\Lambda_{re}^{ra}(t)\right]$ and $E\left[\Lambda_{rs}^{ra}(t)\right]$ are the expected maintenance effort expenditures considering the reporter and assignee until the end of operation. Also, $E\left[\Lambda_{re}^r(t)\right]$ and $E\left[\Lambda_{rs}^r(t)\right]$ means the expected maintenance effort expenditures considering the reporter until the end of operation.

Similarly, the sample path of fixing effort required for OSS maintenance up to operational time t in case of $\Lambda_e(t)$ and $\Lambda_s(t)$ are respectively given by

$$\Theta_e(t) = \Lambda_{re}^r(t) - \Lambda_{re}^{ra}(t), \qquad (2.27)$$

$$\Theta_s(t) = \Lambda_{rs}^r(t) - \Lambda_{rs}^{ra}(t). \qquad (2.28)$$

The zero point of fixing effort $E[\Theta_e(t)]$ and $E[\Theta_s(t)]$ mean the starting point of surplus effort. On the other hand, the zero point of fixing effort $E[\Theta_e(t)]$ and $E[\Theta_s(t)]$ is the end point of effort required in the OSS project. Therefore, the OSS project managers will be able to judge the necessity of fixing effort and stability of OSS from the starting point of surplus effort.

Moreover, we can obtain the CPI by using the following equations:

$$E[CPI_e(t)] = \frac{E\left[\Lambda_{re}^r(t)\right]}{E\left[\Lambda_{re}^{ra}(t)\right]}, \qquad (2.29)$$

$$E[CPI_s(t)] = \frac{E\left[\Lambda_{rs}^r(t)\right]}{E\left[\Lambda_{rs}^{ra}(t)\right]}. \qquad (2.30)$$

Similarly, the sample path of CPI in case of $\Lambda_e(t)$ and $\Lambda_s(t)$ are respectively given by

$$CPI_e(t) = \frac{\Lambda_{re}^r(t)}{\Lambda_{re}^{ra}(t)}, \qquad (2.31)$$

$$CPI_s(t) = \frac{\Lambda_{rs}^r(t)}{\Lambda_{rs}^{ra}(t)}. \qquad (2.32)$$

Furthermore, we can obtain the EAC by using the following equations:

$$E[EAC_e(t)] = E\left[\Lambda_{re}^{ra}(t)\right] + \frac{\left(BAC - E\left[\Lambda_{re}^{r}(t)\right]\right)}{E[CPI_e(t)]}, \qquad (2.33)$$

$$E\left[EAC_s(t)\right] = E\left[\Lambda_{rs}^{ra}(t)\right] + \frac{\left(BAC - E\left[\Lambda_{rs}^{r}(t)\right]\right)}{E[CPI_s(t)]}. \qquad (2.34)$$

Similarly, the sample path of EAC in case of $\Lambda_e(t)$ and $\Lambda_s(t)$ are respectively given by

$$EAC_e(t) = \Lambda_{re}^{ra}(t) + \frac{\left(BAC - \Lambda_{re}^{r}(t)\right)}{CPI_e(t)}, \qquad (2.35)$$

$$EAC_s(t) = \Lambda_{rs}^{ra}(t) + \frac{\left(BAC - \Lambda_{rs}^{r}(t)\right)}{CPI_s(t)}. \qquad (2.36)$$

Finally, we can obtain the ETC by using the following equations:

$$E[ETC_e(t)] = \frac{\left(BAC - E\left[\Lambda_{re}^{r}(t)\right]\right)}{E[CPI_e(t)]}, \qquad (2.37)$$

$$E[ETC_s(t)] = \frac{\left(BAC - E\left[\Lambda_{rs}^{r}(t)\right]\right)}{E[CPI_s(t)]}. \qquad (2.38)$$

Similarly, the sample path of ETC in case of $\Lambda_e(t)$ and $\Lambda_s(t)$ are respectively given by

$$ETC_e(t) = \frac{\left(BAC - \Lambda_{re}^{r}(t)\right)}{CPI_e(t)}, \qquad (2.39)$$

$$ETC_s(t) = \frac{\left(BAC - \Lambda_{rs}^r(t)\right)}{CPI_s(t)}. \qquad (2.40)$$

2.5 NUMERICAL EXAMPLES

This section shows the Apache Tomcat known as the OSS developed under Apache Software Foundation (www.apache.org 2018). The project data used in this chapter is obtained from the bug tracking system on the Apache Software Foundation website. In particular, the effort data of version 7 after the year 2011 are applied in order to understand the maintenance effort in Apache Tomcat.

As an example of the stability assessment for the OSS project, the estimated CPI in case of $E[CPI_s(t)]$ and $CPI_s(t)$ is shown in Figure 2.1 as the assessment measures of EVM for OSS project, respectively. From Figure 2.1, the OSS project is executed in a cost-effective manner, because $E[CPI_s(t)]$ and $CPI_s(t)$ exceeds the value of 1 after 3145.229 days.

Moreover, we show the estimated transition probability distribution in Equations 2.15 and 2.19 in Figures 2.2 and 2.3,

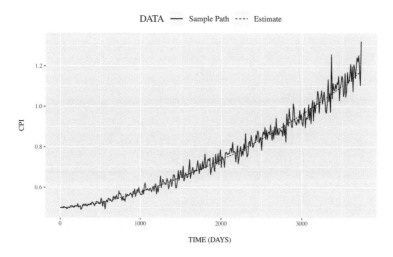

FIGURE 2.1 The estimated CPI in case of $E[CPI_s(t)]$ and $CPI_s(t)$.

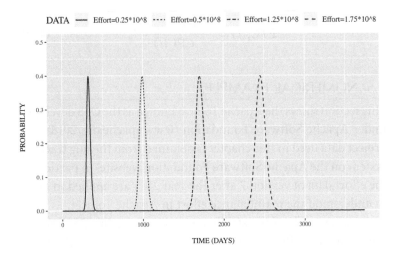

FIGURE 2.2 The estimated transition probability distribution in Equation 2.15.

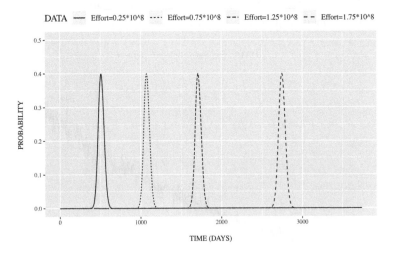

FIGURE 2.3 The estimated transition probability distribution in Equation 2.19.

respectively. From Figures 2.2 and 2.3, we find that the variation becomes large according to the operation procedures in the case of Equation 2.15 in Figure 2.2.

2.6 CONCLUSION

This chapter has discussed the method of earned value analysis considering various external factors associated with OSS project. In the past, several methods of OSS reliability assessment have been proposed. However, it will be important for OSS project managers to control the progress of the OSS project in terms of the OSS management effort. Also, the appropriate control of management effort for OSS will indirectly link to the quality, reliability, and cost reduction of OSS. In terms of OSS management effort, we have proposed the method of earned value analysis based on the stochastic differential equation model and the jump diffusion process model.

In particular, this chapter has proposed the method of earned value analysis for an OSS project by devising the irregular fluctuation from the characteristics of OSS development and management. The proposed method may be helpful as the assessment method of the progress for an OSS project in operation phase. Also, we have found that our method can assess the stability and effort management considering the operational environment of OSS. Moreover, several analyses based on the noisy models have been shown in this chapter. Then, we showed several numerical results based on the characteristics of the Wiener process and the jump term included in our models.

ACKNOWLEDGMENTS

This work was supported in part by the JSPS KAKENHI Grant No. 16K01242 in Japan.

REFERENCES

Arnold, L. "Stochastic differential equations: Theory and applications." 1974.

Fleming, Q. W., and J. M. Koppelman. *Earned Value Project Management* (4th edn.). Project Management Institute, Newton Square, PA, 2010.

Kapur, P. K., H. Pham, A. Gupta, and P. C. Jha. *Software Reliability Assessment with OR Applications.* Springer, London, 2011.

Lyu, M. R. *Handbook of Software Reliability Engineering.* Vol. 222. IEEE Computer Society Press, Los Alamitos, CA, 1996.

Merton, R. C. "Option pricing when underlying stock returns are discontinuous." *Journal of Financial Economics* 3(1–2), 1976: 125–144.

Musa, J. D., A. Iannino, and K. Okumoto. *Software Reliability: Measurement, Prediction, Application.* McGraw-Hill, New York, 1987.

Norris, J. S. "Mission-critical development with open source software: Lessons learned." *IEEE Software* 21(1), 2004: 42–49.

Tamura, Y., H. Sone, and S. Yamada. "Earned value analysis and effort optimization based on Wiener process model for OSS project." *Proceedings of 2018 Asia-Pacific International Symposium on Advanced Reliability and Maintenance Modeling (APARM 2018) and 2018 International Conference on Quality, Reliability, Risk, Maintenance, and Safety Engineering (QR2MSE 2018)*, Qingdao, Shandong, China, 2018, pp. 373–378.

The Apache Software Foundation, The Apache HTTP Server Project, http://www.apache.org/, 2018.

Wong, E. "Stochastic processes in information and dynamical systems." 1971.

Yamada, S. *Software Reliability Modeling: Fundamentals and Applications.* Vol. 5. Springer, Tokyo, 2014.

Yamada, S., M. Kimura, H. Tanaka, and S. Osaki. "Software reliability measurement and assessment with stochastic differential equations." *IEICE Transactions on Fundamentals of Electronics, Communications and Computer Sciences* 77(1), 1994: 109–116.

Yamada, S., and Y. Tamura. *OSS Reliability Measurement and Assessment.* Springer International Publishing, Cham, Switzerland, 2016.

Yamada, S., and Y. Tamura, "Effort management based on jump diffusion model for OSS project." *Proceedings of the 24th ISSAT International Conference on Reliability and Quality in Design*, Toronto, Ontario, Canada, 2018, pp. 126–130.

Zhou, Y., and J. Davis. "Open source software reliability model: An empirical approach." *ACM SIGSOFT Software Engineering Notes* 30(4), 2005, pp. 1–6. ACM.

CHAPTER 3

Software Reliability Modeling for Dynamic Development Environments

N. R. Barraza

CONTENTS

3.1 INTRODUCTION

The last advances in software engineering introduced dynamic processes in software development and testing that improved the classical waterfall methodology. This dynamic environment proposed by the Agile, RUP (Rational Unified Process) and TDD (test-driven development) methodologies among others, demands software reliability models that take into account an iterative coding and testing stages. This is quite different from the large waterfall testing stage where code is constantly introduced in order to fix failures, which leads to the well-known software reliability growth models (SRGMs) mainly based on non-homogeneous Poisson processes. In those mentioned before iterative processes, code is introduced not just in order to fix failures but also to meet new requirements. This situation introduces the concern about an increasing failure rate first stage especially if the software reliability analysis is intended to be applied at the very start of development. The increasing failure rate characteristic makes the SRGM not suitable to be applied at this very first stage where new models and concepts are needed. Following the previously mentioned characteristics, a new model based on a pure birth process is proposed in this chapter. Pure birth processes have been largely applied in software reliability, see for example Okamura and Dohi (2011), and the well-known non-homogeneous Poisson processes are a special type of pure birth processes where the failure (birth) rate is a non-linear function of time. In our formulation, the failure rate is not just a non-linear function of time but also a function of the number of failures previously detected, introducing a contagion process. The concept of contagion arises in order to model the increasing failure rate stage. We describe next the formulation of our model, its assumptions and several applications to real data sets.

3.2 THE WATERFALL AND AGILE METHODOLOGIES

There are six stages in the software life cycle: (1) requirement analysis, (2) design, (3) coding, (4) testing, (5) deployment and (6) maintenance. In the waterfall model, these stages were well separated and used to last a long period of time as shown in Figure 3.1. The first software reliability models developed in the 1970s were intended to model the software failure detection stochastic process executed during the testing phase. Since software failures are constantly removed, we expect reliability growth, that is, a

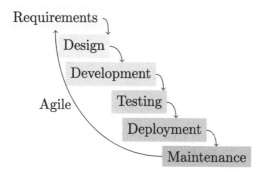

FIGURE 3.1   Software development methodology.

continuous decrease over time in number of failures, then, those models are so-called SRGMs.

The last advances in software engineering introduced improvements in the software development and testing processes. Those improvements have to do with getting efficiency by performing the stages more dynamically, in short iterations as shown in Figure 3.1. Such iterations are commonly referred to as sprints and can last just a week. In this kind of dynamic methodology that comprises, for example Agile, RUP and TDD among others, code is introduced not just to fix failures but to meet new requirements. We can expect as a consequence that failures are not only removed but also introduced with the new code. Then, this scenario cannot be clearly considered as a reliability growth one, and we expect a number of failures over time curve with an increasing failure rate in the first stage, especially when a software reliability analysis is performed at the very first stage. We should focus then on software reliability models that take into account a possible quite long increasing failure rate stage instead of a reliability growth. It is in that sense that we introduce the model presented next. Despite the previous considerations, several approaches on the application of software reliability growth modes to Agile methodologies have been proposed in the literature, see for example Yamada and Kii (2015) and Rawat et al. (2017).

3.3 THE PROPOSED MODEL

Our proposal is based on a pure birth process where the probability of having detected r failures by the time t is given by the following differential equation (Feller 2008):

$$P_r'(t) = -\lambda_r(t)P_r(t) + \lambda_{r-1}(t)P_{r-1}(t) \tag{3.1}$$

Several models may arise depending on the failure rate function $\lambda_r(t)$. The classical software reliability models are based on non-homogeneous Poisson processes like the Goel-Okumoto, Musa-Okumoto and delayed S-shaped models. Non-homogeneous Poisson processes are also pure birth processes where the failure rate is a non-linear function of time (Pham 2010).

The mean value function can be obtained by summing up Equation 3.1 multiplied by r, getting this way the following differential equation for the mean:

$$m'(t) = \sum_{r=1}^{\infty} \lambda_{r-1}(t)P_{r-1}(t) \tag{3.2}$$

Following the considerations exposed in Section 3.2, we proposed a model based on the following assumptions:

- Failures are continuously introduced and removed.

- Code is being constantly added either in order to fix failures and/or to meet new requirements.

- The new added code introduces failures at a rate proportional to the existing number of failures (contagion).

- Because of the failures fixing process, the failure intensity decreases inversely proportional to the elapsed execution time. This assumption is that of the Musa-Okumoto model. Actually, in the Musa-Okumoto model it can be shown that the failure intensity decays exponentially with failures experienced. This factor should be better referred to as Musa-Okumoto like factor.

From the preceding assumptions, we propose the following function for the failure rate:

$$\lambda_r(t) = a\frac{1+br}{1+at} \tag{3.3}$$

The factor $(1 + br)$ in Equation 3.3 corresponds to a contagion process, like that of the Polya stochastic process (see Feller 2008 and Barraza 2016). From our model, contagion may arise when there is an interaction between programmers and testers. By replacing Equation 3.3 in Equation 3.2 we can obtain the mean number of failures:

$$m(t) = (1+at)^b - 1 \tag{3.4}$$

Since the mean number of failures for the Polya contagion process is a linear function of time as it can be demonstrated, our model introduces an improvement in the contagion process for the sake of fitting several data sets. We can see from Equation 3.4 that we can expect two different behaviors whether b is greater or lower than one. In the last case, our model becomes an SRGM.

The mean time between failures (MTTF) can be obtained from the exponential waiting time:

$$P(no\ failures\ on\ T > t - s) = \exp\left(-\int_s^t \lambda_r(t)dt\right) \tag{3.5}$$

This leads to the distribution function:

$$F_T(t;s,r) = \frac{1-\exp\left(-\int_s^t \lambda_r(t)dt\right)}{1-\exp\left(-\int_s^\infty \lambda_r(t)dt\right)} t > s \tag{3.6}$$

The MTTF can be obtained after deriving Equation 3.6 in order to obtain the density function and taking the expectation. For our model, the MTTF after r failures were obtained by the time s results:

$$MTTF(r,s) = \frac{1}{a}\frac{1+as}{br}r = 1,2,3,\ldots \qquad (3.7)$$

3.4 EXPERIMENTS

We next show some data sets in order to show the adjustment of our model and its comparison with other models proposed in the literature.

The fist data set corresponds to a noncritical real time system developed under Agile methodologies. The actual and fitted models are shown in Figure 3.2. As explained earlier, the data exhibits an increasing failure rate stage due to the dynamic development and testing environments.

Predictions from both models are quite different. The DS model predicts an almost immediate reliability growth and the proposed

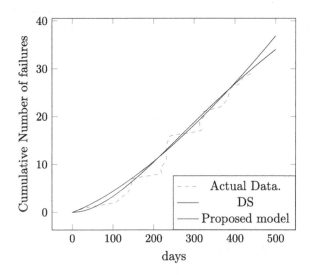

FIGURE 3.2 Agile project fitted models.

model continues with an increasing failure rate. The goodness of fit of the proposed model is slightly better than the DS model.

The second data set corresponds to a client–server system software project developed between 2001 and 2005 and comprises approximately 250 KLOC of C language. The project involved both the server (host) and remote terminals communicated through an X25/IP WAN. The system managed a graphical user interface (GUI) for operations and a relational database management. Development and testing processes were performed under a Agile/waterfall model methodology. The testing process started soon, just 3 months after development, and both stages continued even after release. The actual data and the fitted models are shown in Figure 3.3. It can be seen that our model outperforms the delayed S-shaped model. Performance metrics are evaluated through the predictive ratio risk (PRR; see Pham 2010) and their values are shown in Tables 3.1 and 3.2.

From Table 3.1 we can see that the proposed model performs slightly better than the DS model for the Agile project.

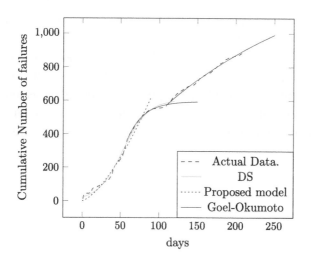

FIGURE 3.3 Mixed Agile/waterfall project fitted models.

TABLE 3.1   Goodness of Fit for the Agile Project

Model	PRR
DS	465.918
Proposed	436.028

TABLE 3.2   Goodness of Fit for the Mixed Agile/
Waterfall Model for the First Stage (First 80 Days)

Model	PRR
Proposed	1398.1
DS	5554.96
Goel-Okumoto	971.329

For the mixed Agile/waterfall model, we can see from Table 3.2 that the proposed model outperforms the DS model and the Goel-Okumoto model performs the best.

3.5 CONCLUSION

A new software reliability model was proposed. It consists of a pure birth process where the failure rate depends not just on time but on the previously detected number of failures. Introducing this way a contagion phenomenon. Two experiments on real data sets with comparisons to other models were presented, where we can confirm that there are no models that perform the best, but some models are more suitable to be applied to a given data set and on a given period of time. The experiments demonstrate that our model can be useful for many data sets with an increasing failure rate at first stage.

REFERENCES

Barraza, N. R. "Software Reliability Modeled on Contagion." *2016 IEEE International Symposium on Software Reliability Engineering Workshops (ISSREW)*, pp. 49–50. IEEE, 2016.

Feller, W. *An Introduction to Probability Theory and Its Applications.* Vol. 2. John Wiley & Sons, 2008.

Okamura, H., and T. Dohi. "Unification of Software Reliability Models using Markovian Arrival Processes." *2011 IEEE 17th Pacific Rim International Symposium on Dependable Computing (PRDC)*, pp. 20–27. IEEE, 2011.

Pham, H. *System Software Reliability,* 1st edn. Springer, 2010.

Rawat, S., N. Goyal, and M. Ram. "Software Reliability Growth Modeling for Agile Software Development." *International Journal of Applied Mathematics and Computer Science* 27(4), 2017: 777–783.

Yamada, S., and R. Kii. "Software Quality Analysis for Agile Development." *2015 4th International Conference on Reliability, Infocom Technologies and Optimization (ICRITO)(Trends and Future Directions)*, pp. 1–5. IEEE, 2015.

CHAPTER 4

Modeling Software Vulnerability Correction/Fixation Process Incorporating Time Lag

J. Kaur, Adarsh Anand, and O. Singh

CONTENTS

4.1 INTRODUCTION

With the world ready to step into Industry 4.0, we are being promised a future where all devices are connected to each other, information is being shared across all technologies and artificial intelligence shall be making decisions with very little input from humans. For this to work, it requires software to have a sound security system. Residual faults in the software at its release time can cause a security issue after release. Faults that can compromise the security of software are generally referred to as vulnerabilities. ISO 27005 defines vulnerability as an asset or multiple assets that can be exploited by some threat(s) (www.iso27001security.com 2018). Vulnerabilities in software may arise due to design flaws, lack of awareness, management issues, time constraint, and so on. There are six stages in the vulnerability life cycle, which starts with the introduction of a vulnerability in a software. The second stage is the detection of the vulnerabilities, followed by the third stage of private exploitation. At the next stage the vulnerability is disclosed to the public, whereby it can be exploited by the general public in the public exploitation phase. The sixth stage is the remedial step wherein security upgrades or patches are released to fix the vulnerability (Joh and Malaiya 2010).

Vulnerability exploitation can cause a lot of damage. For instance, in September 2017, Equifax Inc., a credit reporting agency, announced a major data breach which compromised the personal details (full name, Social Security number, driver license number, address, etc.) of 145.5 million customers. In March 2018, it was further announced that 2.4 million additional customers were impacted. Investigations revealed the cause of the breach to be lack of security reviews and failure to follow general security guidelines (www.lifelock.com 2018). As the situation holds, no software is free of vulnerability. But the upside is that firms are realizing the importance of software security and are taking relevant steps to safeguard themselves from a breach. Many techniques are being deployed to find vulnerabilities in the software and counter them before they are exploited. For example, Microsoft has been using penetration testing on its products Office 365 and Azure and recently started using it on Windows too. Microsoft has employed security experts along with its engineers to create red and blue teams, in which the red team is the in-house hackers and the blue team is the security team working against the attacks by the red team (www.wired.com 2018).

In August 2018, 20 security vulnerabilities were discovered in open source electronic medical records management software (OpenEMR) by the cybersecurity team Project Insecurity. The team informed the organization and patches were released a month later that prevented a possible breach of medical records of 100 million people (www.welivesecurity.com 2018). Extra help and responsible disclosure avoided a massive information breach. Major organizations, such as Facebook, Microsoft, Apple, and the US Air Force, organize bug bounty programs wherein white hat or ethical hackers are called to find bugs or vulnerabilities in software. Handsome rewards are given to those who come up with original vulnerabilities in the software. Such activities have led to a considerable rise in the number of vulnerabilities reported per year. An average of 50 new vulnerabilities were reported each day in 2017, which is a steep rise as compared to 28 being reported per day in 2016 and 17 per day in 2015 (www.secludit.com 2018).

Vulnerabilities can be discovered by the white/black hat hackers, the firm's own detection team, reports of users, researchers, government, and so on. The reported vulnerabilities are verified by the debugging team of the software developer and quick action is taken to handle the vulnerabilities before they are exploited by the hackers. Vulnerability discovery is a complicated process. It can be done manually or through the vulnerability detection kits available online that are usually used by hackers to find vulnerabilities in the software. Static techniques such as lexical analysis, parsing, data flow analysis, and taint analysis, and dynamic techniques like fuzzing, fault injection, and dynamic taint analysis are some of the techniques deployed to find vulnerabilities in software (Jimenez et al. 2009).

It has been observed that the discovery of a particular flaw or vulnerability can lead to the discovery of many more vulnerabilities. It is possible for a single logical error to present itself in the form of multiple vulnerabilities. For instance, CVE-2018-8331 presented as an execute code as well as overflow vulnerability with CVSS 9.3 in Microsoft Excel and Microsoft Office 2016. CVE-2014-6120 with CVSS 10.0 severity allowed remote source execute code vulnerability in five versions of IBM's Rational Appscan source and across nine versions of its security Appscan source (www.cvedetails.com). Thus, discovery of a single vulnerability that presented itself across various versions allowed the discovery vulnerabilities across different versions. Often a cyberattack or detection of a particular category of vulnerability opens up avenues to find more vulnerabilities.

Thus, the vulnerabilities can be categorized as leading vulnerabilities and dependent vulnerabilities. The dependency in vulnerability detection translates to dependency in vulnerability correction. The vulnerability correction process has been also been referred to as a vulnerability fixation process, that is the vulnerabilities are corrected or fixed via software patches released to the public. The time delay between the correction of a leading vulnerability and dependent vulnerabilities also needs to be considered. Due to the large number of vulnerabilities being reported, vulnerability prioritization becomes difficult. Once the vulnerability debugging process is initiated, resources are spent in patch creation. But the patch creation process is not instantaneous, which may lead to a time lag depending on the vulnerability detected. In our current proposal, we shall provide a mathematical model to describe the vulnerability fixation process for different categories of vulnerabilities, namely the leading vulnerabilities and dependent vulnerabilities while considering the time delay.

Numerous mathematical frameworks have been proposed to model the vulnerability discovery process. Anderson's (2002) thermodynamic model was the first vulnerability discovery model (VDM) and was modeled on the lines of a software reliability growth model (SRGM). The Rescorla exponential model and Rescorla linear model proposed by Rescorla (2005) considered the exponential and linear relations to model the vulnerability discovery process. Alhazmi and Malaiya (2005a) proposed a three-phase, S-shaped logistic model to show the rise and fall of the detection rate over the learning phase, testing phase and saturation phase with time. An effort-based model was also proposed by Alhazmi and Malaiya (2005b) to model the effort consumed in terms of resources and budget in finding vulnerabilities. The vulnerability discovery process over multiple versions of the software has also been modeled by Anand et al. (2017b) and Kim et al. (2007). Bhatt et al. (2018) categorized the detected vulnerabilities on the basis of ease of detection and severity. Arora et al. (2008) discussed the optimal policy for vulnerability disclosure.

The most common approach to fixing a vulnerability is to provide patching service. Patches are corrective code released during the operational phase of software usually to add some functionality, to update it or to remove some bugs. Major changes in the source code are dealt in the software upgrades, while patches are the prompt solution to handle the vulnerabilities or faults. Research like the work done by Jiang and Sarkar (2003) and Anand et al. (2017a) highlight the importance of providing patching service in the software and proposed a scheduling policy for the optimal release of software. Das et al. (2015) and Deepika et al. (2016) explored the role of the tester and user in software reliability growth via the patch service. Anand et al. (2018) considered faults and vulnerabilities simultaneously and presented an optimal patch release policy.

In the earlier VDMs, the vulnerability detection process was considered to be independent, that is detection of one vulnerability was independent of the detection of another vulnerability. With growing advancements in the field, it was realized that in fact this was not the case. It has been observed that detection of one vulnerability leads to the detection of many more vulnerabilities along the same lines. The work of Ozment (2007) gave a conceptualization of the scenario and emphasized the need of a VDM that did not consider independence in detection. He highlighted three main reasons for the dependence: the detection of a new vulnerability, a new location to find vulnerability and use of a new tool to detect vulnerability. Bhatt et al. (2017) proposed a mathematical framework to categorize the discovery of main vulnerabilities and additional vulnerabilities wherein detection of a vulnerability may lead to the discovery of the additional vulnerabilities. Their proposal also enumerated the vulnerabilities identified in each category.

In the current work, we shall be working on the lines of Bhatt et al. (2017) to provide a mathematical model to categorize the vulnerabilities as leading vulnerabilities and dependent vulnerabilities, and modeled the time gap between the correction of the leading vulnerability and dependent vulnerabilities. Various mathematical forms for the time lag have been considered to better simulate this situation.

The rest of chapter has been designed as follows: the modeling framework along with the notations and assumptions used is discussed in Section 4.2, model validation and data analysis are in Section 4.3 followed by the conclusion in Section 4.4. A list of references of articles utilized has also been appended.

4.2 MODEL BUILDING

4.2.1 Notations

The notations used ahead are:

N Constant representing total vulnerability content

N_1 Leading vulnerability content

N_2 Dependent vulnerability content

r_1 Vulnerability removal rate of leading vulnerabilities

r_2 Vulnerability removal for dependent vulnerabilities

p Proportion of leading vulnerabilities corrected/fixed

β Constant

$\phi(t)$ Time lag function

$\Omega_R(t)$ Expected number of vulnerabilities corrected/fixed by time t

$\Omega_{R1}(t)$ Expected number of leading vulnerabilities corrected/fixed by time t

$\Omega_{R2}(t)$ Expected number of dependent vulnerabilities corrected/fixed by time t

4.2.2 Assumptions

Some assumptions used in the modeling are as follows:

1. The vulnerability correction process follows the non-homogeneous Poisson process (NHPP).

2. All the detected vulnerabilities can be categorized as either the leading vulnerability or dependent vulnerability.

3. The mean number of leading vulnerabilities in the time interval $(t, t + \Delta(t))$ is proportional to the mean number of remaining leading vulnerabilities.

4. The dependent vulnerability correction lags the leading vulnerability correction process by a delay factor, say $\phi(t)$.

In their work, Bhatt et al. (2017) proposed a framework to model the vulnerability discovery process for leading and additional vulnerabilities as described by the following differential equation:

$$\frac{d\Omega(t)}{dt} = r(N - \Omega(t)) + s\frac{\Omega(t)}{N}(N - \Omega(t)) \qquad (4.1)$$

where $\Omega(t)$ represents the mean number of vulnerabilities detected at time t, N is the total vulnerability content, r is vulnerability detection rate for main vulnerabilities, and s is the rate of discovery of additional vulnerabilities due to the previously discovered vulnerabilities. Analysis of Equation 4.1 shows that the first component $r(N - \Omega(t))$ represents the main vulnerabilities detected and the second component $s(\Omega(t)/(N))(N - \Omega(t))$ corresponds to the additional vulnerabilities detected due to the previously discovered vulnerabilities.

We have used the aforementioned analogy to study the vulnerability correction process when two different categories of vulnerabilities are involved, that is leading and dependent. Since the vulnerability discovery process follows NHPP hence the restructured model, that is the vulnerability correction process would also follow NHPP.

In our proposal, we have modeled the time lag between the correction of the leading vulnerability and the correction of dependent vulnerabilities. The time lag factor in the fixation of vulnerabilities has been modeled using the analogy of fault dependence, and time lag between detection and removal of faults as given by Huang and Lin (2006).

From the second assumption listed earlier we can infer that the total number of vulnerabilities corrected is the sum total of the independent vulnerabilities and the dependent vulnerabilities corrected, that is

$$N = N_1 + N_2 \qquad (4.2)$$

According to the first assumption, the proposed model, that is vulnerability fixation process, is a mean value function of a NHPP. Let $\Omega_R(t)$ represent the mean number of vulnerabilities fixed in time $(t, t + \Delta t)$. The correction process of leading and depending vulnerabilities is also assumed to follow NHPP. Therefore, $\Omega_R(t)$ can be taken to be the superposition of two NHPPs:

$$\Omega_R(t) = \Omega_{R1}(t) + \Omega_{R2}(t) \qquad (4.3)$$

Consequently, we assume the following functional form for the correction of leading vulnerabilities:

$$\frac{d\Omega_{R1}(t)}{dt} = r_1(N_1 - \Omega_{R1}(t)) \tag{4.4}$$

Solving Equation 4.4 under boundary condition, that is at $t = 0$, $\Omega_{R1}(t) = 0$, we have

$$\Omega_{R1}(t) = N_1(1 - \exp(-r_1 t)) \tag{4.5}$$

From the analogy taken from Bhatt et al. (2017), the second component, that is dependent vulnerabilities, corrected inculcating the time lag function can be denoted by:

$$\frac{d\Omega_{R2}(t)}{dt} = r_2(N_2 - \Omega_{R2}(t))\frac{\Omega_{R1}(t - \phi(t))}{N_1} \tag{4.6}$$

where the differential equation represents the dependent vulnerabilities corrected by time t.

Since there is a time gap between correction pertaining to the leading vulnerability and dependent vulnerability, we can assume that rate of correction for the dependent vulnerability is proportional to the correction for the leading vulnerability and lags by a time lag function $\phi(t)$. Thereby the relation between the two separate yet connected processes can be represented as follows:

$$\Omega_R(t) = \Omega_{R1}(t - \phi(t)) \tag{4.7}$$

The time lag function can be influenced by various factors such as testing environment, code complexity, type and severity of vulnerability. The various forms of time lag function have been used to design different vulnerability correction phenomena through certain models as described in the following sections.

Assuming $N_1 = p.N$ and $N_2 = (1 - p)N$, $0 \le p \le 1$; where p is the proportion of vulnerabilities corrected.

4.2.3 Vulnerability Correction Model 1 (VCM 1)

When $\phi(t) = 0$, that means there is no time lag between correction of the leading and dependent vulnerability. Thus, Equation 4.6 becomes

$$\frac{d\Omega_{R2}(t)}{dt} = r_2 (N_2 - \Omega_{R2}(t)) \frac{N_1(1 - \exp(-r_1 t))}{N_1} \qquad (4.8)$$

Making use of Equation 4.3 and solving Equation 4.8 with boundary condition $t = 0$, $\Omega_{R2}(0) = 0$, we get

$$\Omega_R(t) = N \left(1 - p \exp(-r_1 t) - (1 - p) \exp \left[\frac{r_2}{r_1}(1 - \exp(-r_1 t)) - r_2 t \right] \right)$$

$$(4.9)$$

4.2.4 Vulnerability Correction Model 2 (VCM 2)

In this model, we have considered the scenario when the vulnerability fixation is a two-stage process, that is vulnerabilities are first discovered and then corrected. To formulate this we have taken the analogy from the software fault removal process as described by Yamada et al. (1983). The famous *delayed S-shaped model* given by them describes the fault removal phenomena to be a two-stage process. Faults are detected in the first stage and out of these detected faults some are removed in the second stage. In line with this, the vulnerability correction process can be represented as follows.

Let

$$\Omega_R(t) = N(1 - (1 + r_1 t) \exp(-r_1 t)) \qquad (4.10)$$

From Equation 4.7 we have

$$\Omega_R(t) = \Omega_{R1}(t - \phi(t)) = N(1 - \exp(-r_1(t - \phi(t)))) \qquad (4.11)$$

Equating Equations 4.10 and 4.11 we obtain:

$$N(1-(1+r_1 t)\exp(-r_1 t)) = N(1-\exp(-r_1 t + r_1\phi(t)))\quad(4.12)$$

$$\Rightarrow \phi(t) = \left[\frac{1}{r_1}\log(1+r_1 t)\right]\quad(4.13)$$

Equation 4.13 gives a time delay factor for a two-stage vulnerability correction process. Substituting Equation 4.13 in Equation 4.6 we obtain

$$\frac{d\Omega_{R2}(t)}{dt} = r_2(N_2 - \Omega_{R2}(t))\frac{N_1(1-(1+r_1 t)\exp(-r_1 t))}{N_1}\quad(4.14)$$

Making using of Equation 4.3 and solving Equation 4.14 with boundary condition $t = 0$, $\Omega_{R2}(0) = 0$, we get

$$\Omega_R(t) = N\left(1 - p\exp(-r_1 t) - (1-p)\right.$$

$$\left. \exp\left[\frac{2r_2}{r_1}(1-\exp(-r_1 t)) - r_2 t(1+\exp(-r_1 t))\right]\right)\quad(4.15)$$

4.2.5 Vulnerability Correction Model 3 (VCM 3)

Here we have taken the vulnerability fixation to be a three-stage process, that is vulnerabilities are detected, disclosed and then corrected. This analogy has been drawn from the *three-stage Erlang model* (Kapur et al. 1995) for software fault removal process, which describes the fault removal phenomena happening in three stages. Therefore, the time delay function for the present model can be represented as follows.

Let

$$\Omega_R(t) = N\left(1 - \left(1 + r_1 t + \frac{r_1^2 t^2}{2}\right)\exp(-r_1 t)\right)\quad(4.16)$$

Equating Equations 4.11 and 4.16 we obtain

$$N\left[1-\left(1+r_1 t+\frac{r_1^2 t^2}{2}\right)\exp(-r_1 t)\right]=N(1-\exp(-r_1 t+r_1\phi(t)))$$

(4.17)

$$\Rightarrow \phi(t)=\left[\frac{1}{r_1}\log\left(1+r_1 t+\frac{r_1^2 t^2}{2}\right)\right]$$

(4.18)

Using the time delay factor obtained for a three-stage vulnerability fixation process, that is substituting Equation 4.18 in Equation 4.6, we get

$$\frac{d\Omega_{R2}(t)}{dt}=r_2(N_2-\Omega_{R2}(t))\frac{N_1\left[1-\left(1+r_1 t+\frac{r_1^2 t^2}{2}\right)\exp(-r_1 t)\right]}{N_1}$$

(4.19)

Making use of Equation 4.3 and solving Equation 4.19 with boundary condition $t=0$, $\Omega_{R2}(0)=0$, we get

$$\Omega_R(t)=N\left(1-p\exp(-r_1 t)-(1-p)\right.$$

$$\left.\exp\left[\frac{3r_2}{r_1}(1-(1+r_1 t)\exp(-r_1 t))-r_2 t\left(1-\left(1-\frac{r_1 t}{2}\right)\exp(-r_1 t)\right)\right]\right)$$

(4.20)

4.2.6 Vulnerability Correction Model 4 (VCM 4)

In this model, we have taken the vulnerability correction process to follow an S-shaped learning pattern, that is with the passage of time the vulnerability fixation process improves. The *inflection S-shaped model* (Kapur and Garg 1992) from the software fault removal process has been taken to draw an analogy between the vulnerability correction process and fault removal process. The time delay function for such a scenario is obtained as follows. Let

$$\Omega_R(t) = N\left(\frac{1 - \exp(-r_1 t)}{1 + \beta \exp(-r_1 t)}\right) \qquad (4.21)$$

Equating Equations 4.11 and 4.21 we obtain

$$N\left(\frac{1 - \exp(-r_1 t)}{1 + \beta \exp(-r_1 t)}\right) = N(1 - \exp(-r_1 t + r_1 \phi(t))) \qquad (4.22)$$

$$\Rightarrow \phi(t) = \left[\frac{1}{r_1} \log\left(\frac{1 + \beta}{1 + \beta \exp(-r_1 t)}\right)\right] \qquad (4.23)$$

Using the time lag function from Equation 4.23 in Equation 4.6, we get

$$\frac{d\Omega_{R2}(t)}{dt} = r_2(N_2 - \Omega_{R2}(t))\frac{N_1(1 - \exp(-r_1 t))}{N_1(1 + \beta \exp(-r_1 t))} \qquad (4.24)$$

Making use of Equation 4.3 and solving Equation 4.24 with boundary condition $t = 0$, $\Omega_{R2}(0) = 0$, we get

$$\Omega_R(t) = N\left(1 - p\exp(-r_1 t) - (1 - p)\exp(-r_2 t)\right.$$

$$\left.\left[\frac{1 + \beta}{1 + \beta \exp(-r_1 t)}\right]^{\frac{r_2(1+\beta)}{r_1 \beta}}\right) \qquad (4.25)$$

4.3 MODEL ANALYSIS AND DATA VALIDATION

For the purpose of verification, the proposed models have been tested on five data sets. The data sets were extracted from the Common Vulnerabilities and Exposure Database (CVE) (www.cvedetails.com). The vulnerability discovery data was utilized to create the vulnerability fixation data. For this purpose, the vulnerabilities were categorized on the basis of the Common Vulnerability Scoring System, that is the CVSS score. CVSS assigns a score to all vulnerabilities based on numerous factors, which in turn help the firm to gauge the risk that a particular vulnerability poses. The CVSS score can be used to further categorize the vulnerabilities qualitatively. The vulnerabilities with CVSS scores between 0.1 and 3.9 are considered to be *low severity* vulnerabilities; the vulnerabilities with CVSS scores between 4.0 and 6.9 are considered to be *medium severity* vulnerabilities; the vulnerabilities with CVSS scores between 7.0 and 8.9 are considered to be *high severity* vulnerabilities; and the vulnerabilities with CVSS scores between 9.0 and 10 are considered to be *critical severity* vulnerabilities.

The dynamic nature of the IT industry doesn't allow the vendor to keep his product flawless at all times. Constraints such as time, budget, available resources, and market demands require a trade-off between the quality of the software and other factors. Thus, it was assumed that each year the software firm would be patching at least 80% of their vulnerabilities. Further due to the varying levels of risk involved with each category of vulnerability, the percentage of vulnerabilities the firm would be catering to will vary for each category. For the current study we have assumed that all the vulnerabilities of the critical and high severity level would be patched, while 80% of the medium severity and 65% of the low severity category would be handled. The managerial decisions of the firm can influence the percentage levels and thus can be varied as per the firm's requirement.

To show the generic nature of the proposed research work, the models have been analyzed on three different types of products: web browser, operating system and application software. The first and second data sets are of vulnerabilities discovered in web browsers. The first data set (DS-I) comprises 1546 vulnerabilities discovered in the web browser Google Chrome between 2008 and 2018. The second data set (DS-II) comprises 1742 vulnerabilities discovered in the web browser Mozilla Firefox between 2003 and 2018. The third and fourth data sets pertain to vulnerabilities detected in operating systems. The third data set (DS-III) comprises 1116 vulnerabilities discovered in the operating system Microsoft Window Server 2008 between 2007 and 2018, and the fourth data set (DS-IV) comprises 2083 vulnerabilities discovered in the operating system Apple Mac OS X between 1999 and 2018. The fifth data set (DS-V) contains 1052 vulnerabilities discovered in the application software Adobe Flash Player between 2005 and 2018.

The four models have been estimated on the five data sets using the analytical software Statistical Analysis System (SAS 2004). The estimated value of the parameters of the proposed models is shown in Table 4.1. The validity of the results can be verified using

TABLE 4.1   Parameter Estimation

Data Set	Model	N	p	r_1	r_2	β
DS-I	VCM 1	1574.53	0.010	0.068	0.587	—
	VCM 2	1487.76	0.016	0.400	0.345	—
	VCM 3	1594.46	0.013	0.878	0.230	—
	VCM 4	1587.43	0.010	0.792	0.231	10.450
DS-II	VCM 1	2234.14	0.015	0.010	0.800	—
	VCM 2	1759.71	0.128	0.106	0.425	—
	VCM 3	1730.71	0.106	0.246	0.283	—
	VCM 4	2065.22	0.021	0.044	0.597	2.174
DS-III	VCM 1	1027.90	0.015	0.031	0.765	—
	VCM 2	1853.00	0.282	0.063	0.837	—
	VCM 3	1566.77	0.226	0.134	0.921	—
	VCM 4	1317.86	0.105	0.231	0.862	32.956
DS-IV	VCM 1	1908.25	0.008	0.006	0.946	—
	VCM 2	2056.71	0.002	0.038	0.967	—
	VCM 3	2145.71	0.075	0.081	0.891	—
	VCM 4	2087.10	0.001	0.157	0.272	15.640
DS-V	VCM 1	2258.73	0.022	0.007	0.907	—
	VCM 2	3688.48	0.002	0.033	0.941	—
	VCM 3	1903.03	0.010	0.115	0.909	—
	VCM 4	1507.49	0.013	0.218	0.281	11.240

TABLE 4.2   Comparison Criteria

Data Set	Model	SSE	MSE	Root MSE	R-Squared	Adjusted R-Squared
DS-I	VCM 1	18190.0	2021.1	44.9568	0.9923	0.9915
	VCM 2	10847.4	1355.9	36.8228	0.9954	0.9943
	VCM 3	**9096.7**	**1137.1**	**33.7208**	**0.9962**	**0.9952**
	VCM 4	11182.8	1397.9	37.3879	0.9953	0.9941
DS-II	VCM 1	**12915.8**	**993.5**	**31.5202**	**0.9962**	**0.9956**
	VCM 2	15699.1	1207.6	34.7509	0.9954	0.9947
	VCM 3	19880.6	1529.3	39.1059	0.9942	0.9933
	VCM 4	13024.7	1001.9	31.6528	0.9962	0.9956
DS-III	VCM 1	39860.6	3623.7	60.1971	0.9612	0.9612
	VCM 2	**11254.0**	**1250.4**	**35.3616**	**0.9890**	**0.9866**
	VCM 3	13289.0	1476.6	38.4260	0.9871	0.9842
	VCM 4	11491.3	1276.8	35.7324	0.9888	0.9863
DS-IV	VCM 1	551481.0	29025.3	170.4000	0.9001	0.9001
	VCM 2	235917.0	12416.7	111.4000	0.9573	0.9573
	VCM 3	**179745.0**	**9985.8**	**99.9292**	**0.9674**	**0.9656**
	VCM 4	250967.0	13942.6	118.1000	0.9545	0.9520
DS-V	VCM 1	200257.0	15404.4	124.1000	0.9011	0.9011
	VCM 2	66912.2	5576.0	74.6727	0.9670	0.9642
	VCM 3	**55841.1**	**4653.4**	**68.2160**	**0.9724**	**0.9701**
	VCM 4	157094.0	13091.2	114.4000	0.9224	0.9159

different goodness-of-fit measures. Here, five criteria—SSE, MSE, root MSE, R- squared and adjusted R-squared—have been utilized. The calculated values are shown in Table 4.2.

We can see that depending on the data set, different models are a better fit for a particular data set. For instance, VCM 3 gives the best results on data sets DS-I, DS-IV and DS-V, whereas VCM 1 fits better on DS-II, and VCM 2 is a good fit for DS-III. Thus, we can conclude that the proposed models are data specific. The goodness of fit of the models is graphically represented in Figures 4.1–4.5.

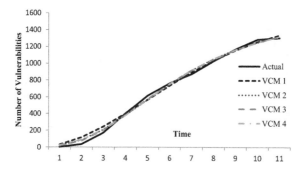

FIGURE 4.1   Goodness of fit of VCMs 1 to 4 on DS-I (Google Chrome).

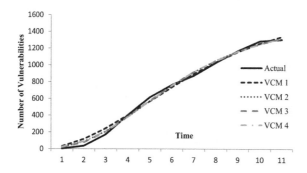

FIGURE 4.2   Goodness of fit of VCMs 1 to 4 on DS-II (Mozilla Firefox).

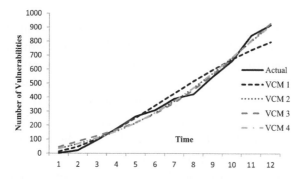

FIGURE 4.3   Goodness of fit of VCMs 1 to 4 on DS-III (Windows Server 2008).

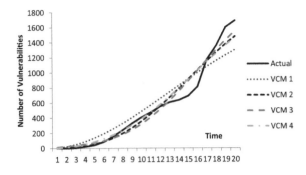

FIGURE 4.4   Goodness of fit of VCMs 1 to 4 on DS-IV (Apple Mac OS X).

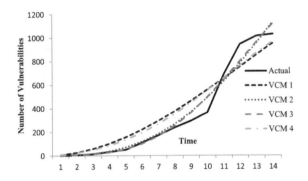

FIGURE 4.5   Goodness of fit of VCMs 1 to 4 on DS-V (Adobe Flash Player).

4.4 CONCLUSION

Security is of prime concern to every software-related firm. Thereby, the need to create models that better simulate the software security scenario plays a crucial role in overall security determination. Software contains vulnerabilities that can be exploited over time. Vulnerability fixation needs to be better understood so their exploitation can be prevented. In the current proposal, we modeled the interdependence in the vulnerabilities, which leads to dependency in the vulnerability fixation. The proposal classifies the vulnerabilities as leading vulnerabilities and dependent vulnerabilities. The time lag between the fixation done for leading vulnerabilities and dependent vulnerabilities has been inculcated in the proposed modeling framework. Varying forms for the time-dependent lag function have been considered, and the four proposed VCMs have been validated on five vulnerability data sets collected from the CVE data set. The results obtained support our claim to dependency in the vulnerabilities.

REFERENCES

Alhazmi, O. H., and Y. K. Malaiya. "Modeling the Vulnerability Discovery Process." *16th IEEE International Symposium on Software Reliability Engineering*, 2005. pp. 10-pp. IEEE, 2005a.

Alhazmi, O. H., and Y. K. Malaiya. "Quantitative Vulnerability Assessment of Systems Software." *Annual Reliability and Maintainability Symposium, 2005. Proceedings.* pp. 615–620. IEEE, 2005b.

Anand, A., M. Agarwal, Y. Tamura, and S. Yamada. "Economic Impact of Software Patching and Optimal Release Scheduling." *Quality and Reliability Engineering International* 33(1), 2017a: 149–157.

Anand, A., P. Gupta, Y. Klochkov, and V. S. S. Yadavalli. "Modeling Software Fault Removal and Vulnerability Detection and Related Patch Release Policy." In: *System Reliability Management*, pp. 35–50. CRC Press, 2018.

Anand, A., S. Das, D. Aggrawal, and Y. Klochkov. "Vulnerability Discovery Modelling for Software with Multi-Versions." In: *Advances in Reliability and System Engineering*, pp. 255–265. Springer, Cham, 2017b.

Anderson, R. "*Security in Open Versus Closed Systems—The Dance of Boltzmann, Coase and Moore*." Technical report, Cambridge University, England, 2002.

Arora, A., R. Telang, and H. Xu. "Optimal Policy for Software Vulnerability Disclosure." *Management Science* 54(4), 2008: 642–656.

Bhatt, N., A. Anand, D. Aggrawal, and O. H. Alhazmi. "Categorization of Vulnerabilities in a Software." In: *System Reliability Management*, pp. 121–135. CRC Press, 2018.

Bhatt, N., A. Anand, V. S. S. Yadavalli, and V. Kumar. "Modeling and Characterizing Software Vulnerabilities." *International Journal of Mathematical, Engineering and Management Sciences (IJMEMS)* 2(4), 2017: 288–299.

Das, S., A. Anand, O. Singh, and J. Singh. "Influence of Patching on Optimal Planning for Software Release and Testing Time." *Communications in Dependability and Quality Management* 18(4), 2015: 81–92.

Deepika, A. A., N. Singh, and D. Pankaj. "Software Reliability Modeling Based on In-House and Field Testing." *Communications in Dependability and Quality Management* 19(1), 2016: 74–84. https://secludit.com/en/blog/spectre-meltdown-vulnerabilities-discovered-2017.

Huang, C.-Y., and C.-T. Lin. "Software Reliability Analysis by Considering Fault Dependency and Debugging Time Lag." *IEEE Transactions on Reliability* 55(3), 2006: 436–450.

Jiang, Z., and S. Sarkar. "Optimal Software Release Time with Patching Considered." *Workshop on Information Technologies and Systems*, Seattle, WA. 2003.

Jimenez, W., A. Mammar, and A. Cavalli. "Software Vulnerabilities, Prevention and Detection Methods: A Review 1." In: *Security in Model-Driven Architecture* 2009, pp. 6–13.

Joh, H., and Y. K. Malaiya. "A Framework for Software Security Risk Evaluation Using the Vulnerability Lifecycle and CVSS Metrics." *Proc. International Workshop on Risk and Trust in Extended Enterprises*, pp. 430–434, 2010.

Kapur, P. K., and R. B. Garg. "A Software Reliability Growth Model for an Error-Removal Phenomenon." *Software Engineering Journal* 7(4), 1992: 291–294.

Kapur, P. K., S. Younes, and S. Agarwala. "Generalized Erlang Software Reliability Growth Model." *ASOR Bulletin* 14(1), 1995: 5–11.

Kim, J., Y. K. Malaiya, and I. Ray. "Vulnerability Discovery in Multi-Version Software Systems." *10th IEEE High Assurance Systems Engineering Symposium, 2007.* pp. 141–148. IEEE, 2007.

Ozment, J. A. "Vulnerability Discovery & Software Security." *PhD diss.*, University of Cambridge, 2007.

Rescorla, E. "Is Finding Security Holes a Good Idea?" *IEEE Security & Privacy* 3(1), 2005: 14–19.

SAS Institute Inc. *SAS/ETS User's Guide Version 9.1.* Cary, NC: SAS Institute Inc., 2004.

www.cvedetails.com.

www.iso27001security.com/html/27005.html.

www.lifelock.com/learn-data-breaches-equifax-data-breach-2017.html.

www.welivesecurity.com/2018/08/08/software-bugs-100-million-health-records-risk-exposure.

www.wired.com/story/microsoft-windows-red-team.

Yamada, S., M. Ohba, and S. Osaki. "S-Shaped Reliability Growth Modeling for Software Error Detection." *IEEE Transactions on Reliability* 32(5), 1983: 475–484.

Change-Point–Based Software Reliability Modeling and Its Application for Software Development Management

Y. Minamino, S. Inoue, and S. Yamada

CONTENTS

5.1 INTRODUCTION

Quality, cost, and delivery (QCD) (Yamada 2011) are important software development management factors that software development managers control during the testing phase of the software development process. In particular, the software development management techniques based on software reliability growth models (SRGMs) (Lin and Huang 2008, Misra 2008, Pham 2008, Teng et al. 2006, Yamada 2011, 2014, Yamada and Osaki 1985, Yamada and Tamura 2016) are often used. The majority of SRGMs are developed by treating the software failure-occurrence time or the fault-detection time intervals as random variables, and they assume that the stochastic characteristics of these quantities are the same throughout the testing phase. However, we often observe that stochastic characteristics of software failure-occurrence time intervals change because of a change in the testing environment. The testing time observed during such a phenomenon is called *change-point*, and it is considered to be a factor that influences the accuracy of software reliability assessment (Huang 2005a, Huang and Lin 2010, Zhao 1993, Inoue and Yamada 2007, 2010, 2011, Chang 2001).

SRGMs with the change-point have been discussed by several researchers (Huang 2005a, Huang and Hung 2010, Shyur 2003, Inoue and Yamada 2007, 2011) who have considered the difference between the software failure-occurrence rates before and after the change-point. However, it is necessary to consider between the software failure-occurrence time intervals before and after the change point since the same software product is being tested even if the testing environment changes during the testing phase. Therefore, we focus on the later relationship and develop new SRGMs that consider the change-point to achieve a higher accuracy in software reliability assessment.

In addition, we propose a theoretical change-point detection method based on the Laplace trend test (Gaudoin 1992, Gokhale et al. 1998). The Laplace trend test is a method for observing the time when the trend of the software reliability growth process changes. More specifically, we confirm the effectiveness of our change-point detection method by comparing the goodness of fit of the existing models and our models with the estimated change-points.

Furthermore, we discuss the *optimal software release problem* (Inoue and Yamada 2008, Osaki 2002) based on the change-point model as an application. It is an optimization problem for estimating the optimal release (shipping) time of software products, and the optimal software release time is often determined by several criteria, such as reliability and cost (Pham 2006). In addition, we estimate not only the optimal software release time but also the optimal change-point providing that the change-point is often determined by the experience and intuition of software development managers.

We also discuss the optimal software release problem based on the multi-attribute utility theory (MAUT) (Singh et al. 2012, 2015, Kapur et al. 2013, 2014). The MAUT is a utility theory on decision-making considering multiple constraints. As an example, when we apply the cost evaluation criterion, the optimal software release time is determined by minimizing the total software cost. However, the minimized cost is not necessarily important, because software development managers determine the optimal software release time by considering multiple evaluation criteria simultaneously. Therefore, we estimate the optimal software release time and the optimal change-point for determining the time of the testing environment change considering multiple constraints.

5.2 CHANGE-POINT MODELING

5.2.1 Introduction

It is known that the change-point occurrence influences the accuracy of software reliability assessment based on SRGMs. Therefore, we will develop new non-homogeneous Poisson process (NHPP) models (Pham 2000, 2008; Lyu 1996) by introducing the concept of the change-point into the existing exponential, delayed S-shaped, and inflection S-shaped SRGMs (Pham 2003). In addition, we introduce the uncertainty of the testing environment into the existing models, when each mean value function follows the exponential and delayed S-shaped SRGMs, and develop a more specific change-point model.

According to previous research, the change-point occurs from the following perspectives in the actual testing phase (Langberg et al. 1985):

- Software development management

  - When software development managers predict that the delivery will be delayed and the reliability requirement will not be achieved

- Technical aspect of software development

  - Due to the difficulty of fault detection, fault independence, and the fault density difference of each module
  - Or due to the skill-up process of test engineers

In the former case, the change-point is determined by software development managers' decisions, such as increasing the test personnel or changing the fault target. On the other hand, in the latter case, the change-point occurs as a natural phenomenon. However, software development managers usually do not have methods for detecting the change-point–based on quantitative grounds.

A change-point detection method using hypothesis testing has been previously proposed and confirmed to have certain effectiveness (Zhao and Wang 2007). However, the proposed method cannot detect four or more change-points. Therefore, we propose a new change-point detection method based on the Laplace trend test. The Laplace trend test is used as a method for observing the time when the trend of change in the software reliability growth process originated. The steps of our approach are as follows. First, we detect four change-points by deriving the Laplace factor from actual data sets. Next, we apply these estimated change-points to the proposed change-point models considering the uncertainty of the testing environment. Then, we compare these change-point models by using mean squared error (MSE) and check the effectiveness of our proposed method.

5.2.2 Change-Point Modeling Framework

The majority of NHPP-based SRGMs, in which the total number of detectable faults is finite, can be developed considering the following basic assumptions (Langberg and Singpurwalla 1985):

A1 Whenever a software failure is observed, the fault that caused it will be detected immediately and no new faults are introduced in the fault-removing activities.

A2 Each software failure occurs at independently and identically distributed random times with the probability distribution $F(t) \equiv \Pr\{T \le t\}$, where $\Pr\{A\}$ represents the probability.

A3 The initial number of faults in a software, $N_0(t \ge 0)$, is a random variable and is finite.

We define that $\{N(t), (t \ge 0)\}$ denotes a counting process representing the total number of faults detected up to the testing time t. Based on the aforementioned basic assumptions, the probability that m faults are detected up to the testing time t is divided as

$$
\begin{aligned}
\Pr\{N(t) = m\} &= \sum_n \binom{n}{m} \{F(t)\}^m \{1 - F(t)\}^{n-m} \frac{\omega^n}{n!} \exp[-\omega] \\
&= \frac{\{\omega F(t)\}^m}{m!} \exp[-\omega F(t)] \quad (m = 0,1,2,\ldots),
\end{aligned}
\tag{5.1}
$$

where the initial fault content is assumed to follow a Poisson distribution with the mean ω. Equation 5.1 is equivalent to an NHPP with the mean value function $E[N(t)] = \omega F(t)]$.

Also, we define the stochastic quantities related to our modeling approach as follows:

X_i: ith software failure-occurrence time before the change-point $(X_0 = 0, i = 0, 1, 2, \ldots)$.

S_i: ith software failure-occurrence time interval before the change-point $(S_i = X_i - X_{i-1}, S_0 = 0, i = 0, 1, 2, \ldots)$.

Y_i: ith software failure-occurrence time after the change-point $(Y_0 = 0, i = 0, 1, 2, \ldots)$.

T_i: ith software failure-occurrence time interval after the change-point $(T_i = Y_i - Y_{i-1}, T_0 = 0, i = 0, 1, 2, \ldots)$.

We assume that the respective relationship for the stochastic quantities before and after the change-point as follows (Okamura et al. 2001):

$$\begin{cases} Y_i = \alpha(X_i), \\ T_i = \alpha(S_i), \\ K_i = J_i(\alpha^{-1}(t)), \end{cases} \tag{5.2}$$

where $\alpha(t)$ represents a testing environment function representing the relationship between the stochastic quantities of the software failure-occurrence time or time intervals before and after the change-point. $J_i(t)$ and $K_i(t)$ are the probability functions related to the random variables $S_i(t)$ and $T_i(t)$, respectively. The testing environment function is assumed as follows:

$$\alpha(t) = \alpha t \quad (\alpha > 0), \tag{5.3}$$

where α is a proportional constant representing the relative magnitude of the effect of the change-point on the software reliability growth process and it is called the environmental factor (Inoue and Yamada 2007, 2008, 2011). $\alpha > 1$ indicates that the software failure-occurrence time interval is longer than the software failure-occurrence time interval before the change-point. It means that it is difficult to detect the faults. On the other hand, $0 < \alpha < 1$ indicates that the software failure-occurrence time interval is shorter than the software failure-occurrence time interval before the change-point. It means that it is easy to detect the faults.

Furthermore, we apply the gamma distribution as a probability function that can express the difference between the software failure-occurrence phenomenon with uncertainty before and after the change-point. The gamma distribution (Lyu 1996) is defined as

$$f_k(\alpha) = \frac{\theta^k \alpha^{k-1} e^{-\theta \alpha}}{\Gamma(k)} \quad (k > 0, \alpha > 0, \theta > 0), \tag{5.4}$$

where $\Gamma(k)$ is the gamma function, and k and θ are the parameters of the gamma distribution. Using Equations 5.3 and 5.4, the testing-environment function is derived as

$$\int_0^\infty \alpha(t) f_k(\alpha) \, d\alpha = \frac{k}{\theta} t. \tag{5.5}$$

Suppose that n faults have been detected up to the change-point and their fault-detection time from the test beginning ($t = 0$) have been observed as $0 < x_1 < x_2 < \cdots < x_n \leq \tau$, where τ represents the change-point. Then, the probability distribution function of T_1, a random variable representing the time interval from the change-point to the first software failure occurrence after the change-point, can be derived as

$$\overline{K_1} \equiv \Pr(T_1 > t) = \frac{\Pr\left\{ S_{n+1} > \tau - x_n + \dfrac{t}{\alpha} \right\}}{\Pr\{S_{n+1} > \tau - x_n\}} \\ = \frac{\exp\left[-\left\{ M_B\left(\tau + \dfrac{t}{\alpha} \right) - M_B(x_n) \right\} \right]}{\exp[-M_B(\tau) - M_B(x_n)]}, \tag{5.6}$$

where $\overline{K_1}(t)$ indicates the co-function of the probability distribution function $K_1(t) \equiv \Pr\{T_1 \leq t\}$, that is, $\overline{K_1} \equiv 1 - K_1(t)$, and $M_B(t) \ (\equiv \omega J_1(t))$ is the expected number of faults detected up to the change-point. From Equation 5.6, the expected number of faults detected up to $t \in (\tau, \infty]$ after the change-point, $M_A(t)$, is formulated as

$$M_A(t) = -\log \Pr\{T_1 > t - \tau\} = \log \overline{K_1}(t - \tau). \tag{5.7}$$

Therefore, the expected number of faults detected after the change-point–based on the testing-environment factor in Equation 5.3 is derived as

$$M_A(t) = M_B\left(\tau + \frac{t-\tau}{\alpha}\right) - M_B(\tau). \qquad (5.8)$$

In addition, the expected number of faults detected after the change-point–based on the testing-environment factor in Equation 5.5 is derived as

$$M_A(t) = M_B\left(\tau + \frac{\theta}{k}\right) - M_B(\tau). \qquad (5.9)$$

From the preceding, we obtain the following expected number of faults detected up to the testing time t ($t \in [0, \infty)$, $0 < \tau < t$) (Minamino et al. 2011, 2016):

$$\Lambda(t) = \begin{cases} \Lambda_B(t) = M_B(t) & (0 \le t \le \tau), \\ \Lambda_A(t) = M_B(\tau) + M_A(t) = M_B\left(\tau + \dfrac{t-\tau}{\alpha}\right) & (\tau < 0), \end{cases}$$

$$(5.10)$$

$$\Lambda(t) = \begin{cases} \Lambda_{gamma_B}(t) = M_B(t) & (0 \le t \le \tau), \\ \Lambda_{gamma_A}(t) = M_B(\tau) + M_A(t) = M_B\left(\tau + \dfrac{\theta(t-\tau)}{k}\right) & (\tau < 0). \end{cases}$$

$$(5.11)$$

Therefore, the existing exponential, delayed S-shaped, inflection S-shaped SRGMs (EXP, DSS, and ISS models) and their respective change-point models (EXP-CP, DSS-CP, and ISS-CP models) are as follows:

$$\Lambda(t) = \begin{cases} \Lambda_B(t) = \omega\{1 - \exp[-bt]\} & (0 \le t \le \tau), \\ \Lambda_A(t) = \omega\left\{1 - \exp\left[-b\left(\tau + \dfrac{t-\tau}{\alpha}\right)\right]\right\} & (\tau < t), \end{cases} \qquad (5.12)$$

$\Lambda(t)$

$$= \begin{cases} \Lambda_B(t) = \omega\left\{1-(1+bt)\exp[-bt]\right\} & (0 \le t \le \tau), \\ \Lambda_A(t) = \omega\left\{1-\left(1+b\left(\tau+\dfrac{t-\tau}{\alpha}\right)\right)\exp\left[-b\left(\tau+\dfrac{t-\tau}{\alpha}\right)\right]\right\} & (\tau < t), \end{cases}$$

(5.13)

$$\Lambda(t) = \begin{cases} \Lambda_B(t) = \omega\left(\dfrac{1-\exp[-bt]}{1+c\cdot\exp[-bt]}\right) & (0 \le t \le \tau), \\[4mm] \Lambda_A(t) = \omega\dfrac{\left(1-\exp\left[-b\left(\tau+\dfrac{t-\tau}{\alpha}\right)\right]\right)}{\left(1+c\cdot\exp\left[-b\left(\tau+\dfrac{t-\tau}{\alpha}\right)\right]\right)} & (\tau < t), \end{cases}$$

(5.14)

where b is the fault detection rate, l is the inflection coefficient, and $c = ((1 - l)/(l))$, $(0 < l \le 1)$. From Equation 5.11, the exponential and delayed S-shaped change-point models with the uncertainty of the testing-environment factor (EXP2-CP and DSS2-CP models) are as follows (Minamino et al. 2014a,b):

$$\Lambda(t) = \begin{cases} \Lambda_{gamma\_B}(t) = \omega\left\{1-\exp[-bt]\right\} & (0 \le t \le \tau), \\ \Lambda_{gamma\_A}(t) = \omega\left\{1-\exp\left[-b\left(\tau+\dfrac{\theta(t-\tau)}{k}\right)\right]\right\} & (\tau < t), \end{cases}$$

(5.15)

$\Lambda(t)$

$$= \begin{cases} \Lambda_{gamma_B}(t) = \omega\left\{1-(1+bt)\exp[-bt]\right\} & (0 \le t \le \tau), \\ \Lambda_{gamma_A}(t) = \omega\left\{1-\left(1+b\left(\tau+\dfrac{\theta(t-\tau)}{k}\right)\right)\exp\left[-b\left(\tau+\dfrac{\theta(t-\tau)}{k}\right)\right]\right\} \\ \hspace{8cm} (\tau < t). \end{cases}$$

(5.16)

5.2.3 Parameter Estimation and Software Reliability Assessment Measures

We use the maximum likelihood method (Yamada 2014) to estimate the parameters for each model. Suppose that we have observed K data pairs (t_k, y_k), where $(k = 0, 1, 1, \ldots, K)$, and the log-likelihood function is given as

$$
\ln L(\boldsymbol{\theta}\,|\,\tau,\alpha) = \sum_{k=1}^{K}(y_k - y_{k-1})\ln\left[H(t_k;\boldsymbol{\theta}\,|\,\tau,\alpha) - H(t_{k-1};\boldsymbol{\theta}\,|\,\tau,\alpha)\right]
$$
$$
-H_i(t_K;\boldsymbol{\theta}\,|\,\tau,\alpha) - \sum_{k=1}^{K}\ln[(y_k - y_{k-1})!],
$$

$$(5.17)$$

where y_k is the total number of faults detected during $(0, t_k]$ $(0 < t_1 < t_2 < \cdots < t_K)$. $\boldsymbol{\theta}$ is the set of parameters of the model. The estimated parameters are obtained by solving the following log-likelihood equation:

$$
\frac{\partial(\boldsymbol{\theta}\,|\,\tau,\alpha)}{\partial\boldsymbol{\theta}} = 0
$$

$$(5.18)$$

with respect to the parameters of the model.

Software reliability assessment measures based on the SRGMs play an important role in quantitative software reliability assessment. In particular, the expected number of remaining faults and the software reliability function are often used. The expected number of remaining faults by the arbitrary testing time t is formulated as

$$
M(t) \equiv E\left[\bar{N}(t)\right] = E[N(\infty) - N(t)]
$$
$$
= \Lambda(\infty) - \Lambda(t).
$$

$$(5.19)$$

The software reliability function is defined as the probability that a software failure does not occur in the time interval $(t, t + x]$, where $(t \geq 0, x \geq 0)$, given that the testing or operation is executed at testing time t. If the counting process $\{N(t), t \geq 0\}$ follows an NHPP with the mean value function $\Lambda(t)$, the software reliability function is formulated as

$$R(x \mid t) = \exp[-\Lambda(t+x) - \Lambda(t)]. \qquad (5.20)$$

It is noted that Equation 5.20 is derived under the condition that the software system operates in the same environment as the one used in the testing phase after the change-point.

5.2.4 Change-Point Detection Method Based on the Laplace Trend Test

We conducted the Laplace trend test as a change-point detection method. The Laplace factor $l(k)$ is derived as (Minamino et al. 2014a,b)

$$l(k) = \frac{\sum_{i=0}^{k}(i-1)n(i) - \frac{(k-1)}{2}\sum_{i=0}^{k}n(i)}{\sqrt{\frac{k^2-1}{12}\sum_{i=0}^{k}n(i)}}, \qquad (5.21)$$

where $n(i)$ is the number of faults observed during the unit time i. The Laplace factor is interpreted as follows (Minamino et al. 2014a,b):

1. A negative value indicates the decreasing failure intensity, and thus reliability increases.

2. A positive value indicates the increasing failure intensity, and thus reliability decreases.

3. A value between -2 and $+2$ indicates stable reliability.

5.2.5 Numerical Examples

We show numerical examples using the following actual data sets:

DS1: (t_k, y_k) $(k = 1, 2, \ldots, 26; t_{26} = 26, y_{26} = 40, \tau = 18)$

DS2: (t_k, y_k) $(k = 1, 2, \ldots, 29; t_{29} = 29, y_{29} = 73, \tau = 24)$

DS3: (t_k, y_k) $(k = 1, 2, \ldots, 26; t_{26} = 26, y_{26} = 34, \tau = 17)$

DS4: (t_k, y_k) $(k = 1, 2, \ldots, 29; t_{29} = 29, y_{29} = 25, \tau = 18)$

DS5: (t_k, y_k) $(k = 1, 2, \ldots, 28; t_{28} = 28, y_{28} = 43, \tau = 18)$

where t_k is measured in days, and y_k is the total number of faults detected during $[0, t_k]$. These data sets were collected from the actual web-system testing phases. DS1 and DS2 are the S-shaped software reliability growth curved data, and DS3, DS4, and DS5 are the exponential software reliability growth curved data. The change-point was determined as changing or increasing the test personnel by software development managers.

We obtained the estimated parameters of the EXP-CP model for DS5 as follows: $\hat{\omega} = 44.82, \hat{b} = 0.09$, and $\hat{\alpha} = 0.58$. The expected number of remaining faults and software reliability were estimated as follows: $\hat{M}(28.0) \approx 1.82 \approx 2, \hat{R}(1.0 \mid 28.0) \approx 0.77$. Next, we estimated the parameters of the DSS-CP model for DS3 as follows: $\hat{\omega} = 35.34, \hat{b} = 0.16$, and $\hat{\alpha} = 0.64$. The expected number of remaining faults and software reliability were estimated as follows: $\hat{M}(26.0) \approx 1.34 \approx 2, \hat{R}(1.0 \mid 26.0) \approx 0.77$. Finally, we obtained the estimated parameters of the ISS-CP model for DS1 as follows: $\hat{\omega} = 40.23, \hat{b} = 0.29$, and $\hat{l} = 0.02, \hat{\alpha} = 0.63$. The expected number of remaining faults and software reliability were estimated as follows: $\hat{M}(26.0) \approx 0.23 \approx 1, \hat{R}(1.0 \mid 26.0) \approx 0.91$. Figure 5.1 shows the behavior of the estimated DSS-CP model and its 95% confidence limits. From this figure, the estimated software reliability growth curve changes at the change-point along with the actual behavior.

Table 5.1 is the result of the goodness-of-fit comparison of the change-point models with the corresponding existing models based in terms of the MSE. From the table, we can say that the change-point models have a better performance in software reliability assessment compared with the existing models. That is, the effectiveness of considering the change-point was confirmed.

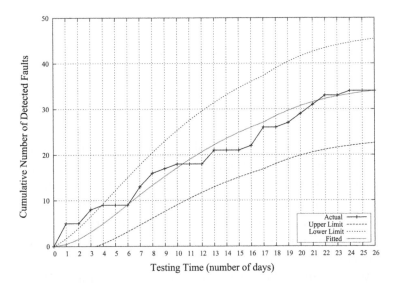

FIGURE 5.1   The estimated DSS-CP model and its 95% confidence limits. (DS3, $\tau = 17$, $\hat{\alpha} = 0.64$, $\hat{\omega} = 35.34$, $\hat{b} = 0.16$). (Adapted from Minamino, Y. et al., "NHPP Models for Software Reliability Measurement with Change-Point," *Proceedings of the 17th ISSAT International Conference on Reliability and Quality in Design*, Vancouver, Canada, August 4–6, 2011, pp. 132–136; Minamino, Y. et al., *Annals of Operations Research* 244(1), 2016: 85–101.)

TABLE 5.1   Comparison of Goodness of Fit

	EXP	EXP-CP	DSS	DSS-CP	ISS	ISS-CP
DS1	**16.08**	20.57	**7.22**	7.78	2.05	*1.78*
DS2	32.08	**24.63**	*19.32*	22.13	**24.35**	31.71
DS3	2.40	*2.19*	6.88	**6.06**	**2.39**	25.80
DS4	0.35	*0.33*	1.70	**1.61**	0.37	**0.38**
DS5	2.47	*2.29*	7.63	**7.50**	2.45	**2.32**

Source: Adapted from Minamino, Y. et al., "NHPP Models for Software Reliability Measurement with Change-Point," *Proceedings of the 17th ISSAT International Conference on Reliability and Quality in Design*, Vancouver, Canada, August 4–6, 2011, pp. 132–136; Minamino, Y. et al., *Annals of Operations Research* 244(1), 2016: 85–101.

On the other hand, we cannot say that the ISS-CP model has a better performance, because the values of the MSE for DS2 and DS3 are large. We suppose that the estimated inflection coefficient might affect the accuracy for the goodness of fit, because the inflection coefficient could not be estimated with another parameter simultaneously.

Figure 5.2 shows the behavior of the Laplace factor for DS1 as a numerical example. From this figure, we can see that the Laplace factor increases after the testing time, $\tau = 4, 7, 9, 16$. The parameters of our EXP2-CP and DSS2-CP models, $\hat{\omega}, \hat{b}, \hat{\theta}$, and $\hat{k}$, in Table 5.2, were estimated using the maximum likelihood method. Table 5.3 shows the result of goodness-of-fit comparison of change-point models with the corresponding existing models

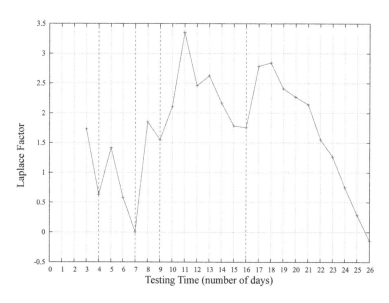

FIGURE 5.2 The temporal behavior of Laplace factor (DS1). (Adapted from Minamino, Y. et al., *Asia-Pacific Journal of Industrial Management (APJIM)*, 5(1), 2014: 63–70; Minamino, Y. et al., "Change-Point Modeling and Detection Method for Software Reliability Assessment," *Proceedings of the 12th International Conference on Industrial Management (ICIM)*, Chendu, China, September 3–5, 2014, pp. 266–270.)

TABLE 5.2   Parameter Estimation Results

	Parameters	$\tau = 4$	$\tau = 7$	$\tau = 9$	$\tau = 16$
EXP2-CP	$\hat{\omega}$	87.70	59.93	45.05	40.22
	$\hat{b}$	0.012	0.021	0.024	0.052
	$\hat{\theta}$	7.04×10^5	3.08×10^6	3.23×10^5	1.60×10^6
	$\hat{k}$	3.29×10^5	1.32×10^6	1.69×10^6	1.88×10^5
DSS2-CP	$\hat{\omega}$	46.47	42.93	42.42	40.12
	$\hat{b}$	0.070	0.074	0.094	0.12
	$\hat{\theta}$	-2.03×10^6	7.19×10^5	1.77×10^6	5.93×10^6
	$\hat{k}$	-9.82×10^5	2.61×10^5	7.50×10^5	1.23×10^6

Source: Adapted from Minamino, Y. et al., *Asia-Pacific Journal of Industrial Management (APJIM)*, 5(1), 2014: 63–70; Minamino, Y. et al., "Change-Point Modeling and Detection Method for Software Reliability Assessment," *Proceedings of the 12th International Conference on Industrial Management (ICIM)*, Chendu, China, September 3–5, 2014, pp. 266–270.

in terms of the MSE. From Table 5.3, we can say that our proposed models with the detected change-points demonstrate better fitting performance compared with the existing models.

5.2.6 Conclusion

We discussed several change-point modeling frameworks for the accuracy improvement of software reliability assessment technique. More specifically, we proposed new change-point models based on the exponential, delayed S-shaped, and inflection S-shaped SRGMs. Also, we discussed the case when the testing-environment factor follows the gamma distribution. In addition, the goodness of fit to the actual data sets was compared by using the MSE, and the performance of our proposed models and the existing models were checked. From the numerical examples, our approach enables us to describe the effect of the change-points on the software reliability growth process.

Furthermore, we discussed a change-point detection method based on the Laplace trend test and checked its effectiveness using

TABLE 5.3 Comparison of Goodness of Fit

DS1			DS2			DS3			DS4			DS5		
τ	EXP2-CP	DSS2-CP	τ	EXP2-CP	DSS2-CP	τ	EXP2-CP	DSS2-CP	τ	EXP2-CP	DSS2-CP	τ	EXP2-CP	DSS2-CP
4	10.95	4.53	4	18.04	19.32	4	2.33	4.56	4	0.43	0.62	4	2.47	3.69
7	7.88	2.92	7	22.92	19.38	7	2.27	6.98	7	0.64	1.72	7	2.75	3.58
9	3.81	2.67	9	29.24	17.13	9	1.67	6.50	9	0.35	1.64	9	5.76	6.29
16	11.80	2.84	16	28.88	16.45	16	2.36	5.17	16	0.26	1.32	16	2.48	7.71
τ	Existing EXP-CP		τ	Existing EXP-CP		τ	Existing EXP-CP		τ	Existing EXP-CP		τ	Existing EXP-CP	
18	14.72		24	26.49		17	2.19		18	0.35		18	4.81	

Source: Adapted from Minamino, Y. et al., *Asia-Pacific Journal of Industrial Management (APJIM)*, 5(1), 2014: 63–70; Minamino, Y. et al., "Change-Point Modeling and Detection Method for Software Reliability Assessment," *Proceedings of the 12th International Conference on Industrial Management (ICIM)*, Chendu, China, September 3–5, 2014, pp. 266–270.

the MSE. On the basis of the numerical examples, we can say that our proposed method enables us to detect the change-points accurately since our models with the detected change-points have better fitting performance.

5.3 OPTIMAL SOFTWARE RELEASE PROBLEM BASED ON THE CHANGE-POINT MODEL

5.3.1 Introduction

Software faults introduced during the development process of a software product are detected and removed using an enormous amount of testing resources during the testing phase. After that, the software product is released to the users. Normally, the more software is tested, the more faults are detected. However, it is difficult to detect all faults because of the budget and delivery date constraints. Therefore, software development managers need to know whether a software product can be shipped without further testing. If some faults are not detected and removed in the testing phase, the remaining faults are detected in the operational phase. It is said that the maintenance cost of modifications and removals in the operational phase is higher than the testing cost. It is also said that the ratio of the debugging costs in each development phase, such as specification, testing, and operation in the software life cycle, is 1:30:140 (Yamada 1994). Therefore, software development managers look to reduce the maintenance cost. However, if the testing time increases to reduce the maintenance cost, the total software cost is expected to increase, and the reliability is expected to increase. On the other hand, if the testing time decreases, the reliability is expected to decrease, and the total software cost is expected to decrease. That is, there is a trade-off between the testing cost and the maintenance cost (Huang 2005b). Furthermore, if the change-point is introduced intentionally in the testing process by software development managers, better timing should be established for appropriate software development management.

Hence, it is important for software development managers to determine when to stop testing and release a software to the user ensuring its adequate reliability (Wang and Pham 2006). This problem is called the *optimal software release problem*. In this section, we discuss the cost-optimal software release problem based on our change-point model using an analytical approach. The steps of our approach are as follows. First, we formulate a cost function based on our proposed EXP-CP model. Then, we consider the relationship of the optimal software release time and the optimal change-point and obtain them simultaneously by minimizing the cost function. In addition, we derive the expected total software cost and the optimal software release policy. Finally, we evaluate the optimal software release time, the optimal change-point, and the expected total software cost using a reliability evaluation criterion.

5.3.2 Optimal Software Release Time and the Change-Point

We discuss the optimal software release problem and the optimal change-point–based on the EXP-CP model according to Equation 5.12. First, we define the following cost parameters:

c_1: Debugging cost per fault before the change-point in the testing phase $(0 < c_1)$

c_2: Debugging cost per fault after the change-point in the testing phase $(0 < c_2)$

c_3: Debugging cost per fault after the change-point in the operational phase $(c_1 < c_3, c_2 < c_3)$

c_4: Testing cost at an arbitrary testing time $(0 < c_4)$

We define that T is the termination time of testing and s is the testing duration from the change-point to the termination time. T^* represents the optimal software release time, and s^* $(0 < s^* < T^*)$ is the optimal testing duration from the change-point to the testing termination time. That is, the optimal change-point is calculated as $\tau^* = T^* - s^*$. From this we obtain the following expected total software cost during the testing and operational phases:

$$C(T,s) = c_1\Lambda_B(T-s) + c_2\{\Lambda_A(T) - \Lambda_B(T-s)\} + c_3\{\omega - \Lambda_A(T)\} + c_4 T, \tag{5.22}$$

where $\Lambda_B(t)$ and $\Lambda_A(t)$ are the mean value functions before and after the change-point, respectively, and we apply the EXP-CP model according to Equation 5.12. Therefore, the optimal software release (termination) time and the optimal testing duration from the change-point to the testing termination time are obtained by solving the following equation analytically:

$$\frac{\partial C(T,s)}{\partial T} = \frac{\partial C(T,s)}{\partial s} = 0. \tag{5.23}$$

From Equation 5.23, we obtain the optimal software release time T^* as

$$T^* \equiv T(s) = \frac{1}{b}\log\left[\frac{\omega b(c_3 - c_2)}{\alpha c_4 \exp\left[\dfrac{bs(1-\alpha)}{\alpha}\right]}\right]. \tag{5.24}$$

Then, the sufficient condition for $T^* > 0$ is given as

$$s < \frac{\log\left[\dfrac{\omega b(c_3 - c_2)}{\alpha c_4}\right]}{b\left(\dfrac{1}{\alpha} - 1\right)} (\equiv A). \tag{5.25}$$

Note that Equation 5.25 is not a constraint condition for the optimization problem. Next, by substituting Equation 5.24 into Equation 5.23, we obtain the following equation:

$$Z(s) \equiv -\alpha c_4 \left[\frac{c_2 - c_1}{c_3 - c_2} \exp\left[\frac{bs}{\alpha}\right] + 1 - \frac{1}{\alpha} \right], \qquad (5.26)$$

where $Z(s)$ is a monotonically decreasing function with respect to s. Furthermore, we can derive the optimal testing duration from the change-point to the testing termination time as

$$s^* = \frac{\alpha}{b} \log\left[\frac{(c_3 - c_2)(1 - \alpha)}{\alpha(c_2 - c_1)} \right]. \qquad (5.27)$$

We can see that when $c_1 < c_2$ and $\alpha < 1$, $Z(s)$ becomes a monotonically decreasing function with respect to s. On the other hand, when $c_2 < c_1$ and $\alpha > 1$, $Z(s)$ becomes a monotonically increasing function with respect to s. Therefore, whether the optimal testing duration from the change-point to the testing termination time exists depends on the relationship between $Z(0)$ and $Z(A)$. From the preceding, we obtain the following cost-optimal software release policy (Minamino et al. 2013, 2016).

I. If $c_1 < c_2$ and $\alpha < 1$, $Z(s)$ becomes a monotonically decreasing function with respect to s.

    1. If $Z(0) > 0$ and $Z(A) < 0$, there is a unique solution. The optimal software release time T^* and the testing duration from the change-point to the testing termination time s^* are given by Equations 5.24 and 5.27, respectively.

    2. If $Z(0) \leq 0$ and $Z(A) < 0$, the testing in the environment before the change-point is continued. The optimal software release time T^* is given by Equation 5.24.

II. If $c_2 < c_1$ and $\alpha > 1$, $Z(s)$ becomes a monotonically increasing function with respect to s.

    1. If $Z(0) < 0$ and $Z(A) > 0$, there is a unique solution. The optimal software release time T^* and the testing duration

from the change-point to the testing termination time s^* are given by Equations 5.24 and 5.27, respectively.

2. If $Z(0) \geq 0$ and $Z(A) > 0$, the testing in the environment before the change-point is continued. The optimal software release time T^* is given by Equation 5.24.

5.3.3 Numerical Examples

We demonstrate an application of the optimal software release policy by using the actual data sets from Section 5.2.5. The cost parameters are $c_1 = 1.0$, $c_2 = 2.0$, $c_3 = 150.0$, and $c_4 = 5.0$.

DS1: From $\hat{\omega} = 4688.87$, $\hat{b} = 0.00039$, $\hat{\alpha} = 1.81\,(> 1)$, and $c_1 < c_2$, the cost-optimal software release policy is not required.

DS2: From $\hat{\omega} = 349.14$, $\hat{b} = 0.0096$, $\hat{\alpha} = 13.29\,(> 1)$, and $c_1 < c_2$, the cost-optimal software release policy is not required.

DS3: From $\hat{\omega} = 44.48$, $\hat{b} = 0.051$, $\hat{\alpha} = 0.78\,(< 1)$, and $c_1 < c_2$, we obtain

$$\begin{cases} s^* = 57.39, \\ Z(A) = -1.75 \times 10^7, \\ Z(0) = 1.07. \end{cases} \tag{5.28}$$

When we apply the cost-optimal software release policy I, the optimal software release time and the expected total software costs are obtained as follows:

$$\begin{cases} T^* = 71.64, \\ C(T^*, s^*) = 501.31, \end{cases} \tag{5.29}$$

DS4: From $\hat{\omega} = 81.02$, $\hat{b} = 0.011$, $\hat{\alpha} = 0.77\,(< 1)$, and $c_1 < c_2$, we obtain

$$\begin{cases} s^* = 256.10, \\ Z(A) = -1.12 \times 10^5, \\ Z(0) = 1.14. \end{cases} \tag{5.30}$$

When we apply the cost-optimal software release policy I, the optimal software release time and the expected total software cost are obtained as follows:

$$T^* = 235.36. \tag{5.31}$$

Note that we cannot say that the optimal solutions are derived properly since s^* is larger than T^*.

DS5: From $\hat{\omega} = 44.82$, $\hat{b} = 0.091$, $\hat{\alpha} = 0.58$ (<1), and $c_1 < c_2$, we obtain

$$\begin{cases} s^* = 29.83, \\ Z(A) = -6.25 \times 10^3, \\ Z(0) = 2.09. \end{cases} \tag{5.32}$$

When we apply the cost-optimal software release policy I, the optimal software release time and the expected total software cost are obtained as follows:

$$\begin{cases} T^* = 37.09, \\ C(T^*, s^*) = 285.36. \end{cases} \tag{5.33}$$

In addition, we determine the optimal software release time based on the cost and reliability evaluation criteria that achieves the reliability objective and minimizes the total software cost. In particular, we set $x = 1.0$ for Equation 5.20 and we assume that the reliability objective is 0.8. We show Figure 5.3 as a numerical example. From Figure 5.3, we can see that the optimal software release time estimated on the basis of the cost and reliability evaluation criteria achieves the reliability objective of 0.8. Therefore, the optimal software release time, the optimal testing duration from the change-point to the testing termination time, and the expected total software cost are also determined in terms of the reliability objective.

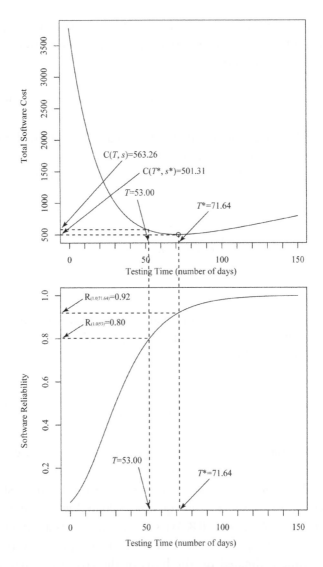

FIGURE 5.3 The optimal total testing period (DS3). (Adapted from Minamino, Y. et al., "NHPP Models with Change-Point for Software Reliability Assessment and its Application to an Optimal Software Release Problem," *Proceedings of the 19th ISSAT International Conference on Reliability and Quality in Design*, Honolulu, Hawaii, August 5–7, 2013, pp. 93–97; Minamino, Y. et al., *Annals of Operations Research* 244(1), 2016: 85–101.)

5.3.4 Conclusion

We discussed the optimal software release problem based on a change-point model and provided the optimal software release policy. In particular, the optimal software release time, the optimal testing duration from the change-point to the testing termination time (the optimal change-point), and the expected total software cost were derived analytically. In addition, we evaluated the optimal software release time by considering a reliability evaluation criterion. Our approach expands the existing optimal software release problem, and we proposed a new method for determining the optimal change-point.

5.4 OPTIMAL SOFTWARE RELEASE PROBLEM BASED ON THE MULTI-ATTRIBUTE UTILITY THEORY

5.4.1 Introduction

We often obtain the optimal software release time as the time minimizing the total software cost. However, minimizing the total software cost is not the most important factor among many other factors (attributes) involved in software development, such as cost and reliability. The important thing for determining the optimal software release time is to consider the various factors together. Therefore, we apply the MAUT, which is a method for decision-making considering the utility, and it enables us to derive and evaluate optimal solutions in terms of the utility theory. The steps of the MAUT are as follows (Minamino et al. 2015a,b). First, we define a cost function based on the proposed EXP-CP and DSS-CP models according to Equations 5.12 and 5.13. Next, we consider the cost and reliability attributes and develop single-attribute utility functions for each attribute. Finally, we define a multi-attribute utility function based on the weighted single-attribute utility functions and maximize it. Then, we can obtain the optimal software release time and the testing duration from the change-point to the testing termination time, simultaneously. The optimal change-point and the expected total software cost are obtained using these results, respectively. In addition, we check the behavior of the multi-attribute utility function and the expected total software cost.

5.4.2 Multi-Attribute Utility Function

First, we assume the cost attribute C as follows:

$$\min : C = \frac{C(T,s)}{C_B}, \tag{5.34}$$

where C_B is the budget and $C(T, s)$ is the cost function shown in Equation 5.22. Then, we assume that the mean value functions, $\Lambda_B(T)$ and $\Lambda_A(T)$, in Equation 5.22 are the EXP-CP and DSS-CP models shown in Equations 5.12 and 5.13. It is desirable to minimize the cost, because software development managers do not want to spend more money than necessary. Therefore, the cost attribute is minimized. Furthermore, we assume the reliability attribute R as follows:

$$\max : R = \frac{\Lambda(t)}{\omega}, \tag{5.35}$$

where the mean value function $\Lambda(t)$ and ω represents the initial fault content. We apply the EXP-CP or DSS-CP models in this section. It is desirable to maximize the reliability attribute, because software development managers look to develop software products with high reliability. Therefore, the reliability attribute is maximized.

Next, we assume that the single-attribute utility functions for each attribute follow the following management strategy:

1. For the cost attribute, at least 50% of the budget must be consumed.

2. For the reliability attribute, at least 50% of software faults should be detected.

3. The management team takes the risk neutral position for each attribute.

Then, the lowest and the highest consumptions for the cost attribute are $C^L = 0.5$ and $C^H = 1.0$, and the lowest and the highest requirements for the reliability attribute are $R^L = 0.5$ and $R^H = 1.0$. We assume the additive linear form because of the risk neutral position in the management strategy. Therefore, we obtain the following single-attribute utility functions:

$$u(C) = 2R - 1, \tag{5.36}$$

$$u(R) = 2C - 1. \tag{5.37}$$

The multi-attribute utility function with some constraints as an optimization problem is given as

$$\begin{aligned} \max: u(C, R) &= W_R \times u(R) - W_C \times u(C) \\ &= W_R \times (2R - 1) - W_C \times (2C - 1) \\ \text{subject to: } W_R + W_C &= 1, \end{aligned} \tag{5.38}$$

where w_C and w_R are weight parameters for each attribute, and $u(C)$ and $u(R)$ are the single-attribute utility functions for each attribute, respectively. The optimal testing time T^* and optimal testing duration from the change-point to the testing termination time s^* are obtained by solving the optimization problem in Equation 5.38. We calculated them using R, which is a tool for statistical analysis.

5.4.3 Numerical Examples

We used DS3 and DS5 from the actual data sets listed in Section 5.2.5. The cost parameters were set to $c_1 = 1.0$, $c_2 = 2.0$, $c_3 = 150.0$, and $c_4 = 5.0$ for both data sets, and the budget was set at $C_B = 1000$ for DS3 and $C_B = 450$ for DS5. Tables 5.4–5.6 show the results of the sensitivity analysis. In addition, we show the results of the estimated optimal change-point and the expected total software cost in these tables. From Tables 5.4 and 5.5, we can see that when the optimal software release time and the optimal testing duration from the change-point to the testing termination time increase, the expected total software cost and the utility also increase. From Table 5.6, we can see that the utility is mainly changed by the weight parameters, although the sensitivity is low. Note that these values are evaluated using comparative assessment.

TABLE 5.4   Results of Sensitivity Analysis (EXP-CP Model, DS3)

w_C	w_R	T^*	s^*	τ^*	$C(T^*, s^*)$	Utility
0.9	0.1	71.90	57.65	14.25	501.33	0.10
0.7	0.3	72.61	58.37	14.25	501.46	0.29
0.5	0.5	73.82	59.58	14.24	502.05	0.49
0.3	0.7	76.32	62.07	14.25	504.53	0.69
0.1	0.9	84.86	70.66	14.20	523.07	0.89

TABLE 5.5   Results of Sensitivity Analysis (EXP-CP Model, DS5)

w_C	w_R	T^*	s^*	τ^*	$C(T^*, s^*)$	Utility
0.9	0.1	37.14	29.87	7.27	285.34	−0.14
0.7	0.3	37.27	30.01	7.26	285.35	0.11
0.5	0.5	37.51	30.25	7.26	285.40	0.36
0.3	0.7	38.03	30.77	7.26	285.66	0.1
0.1	0.9	40.12	32.87	7.25	288.43	0.87

TABLE 5.6   Results of Sensitivity Analysis (DSS-CP Model, DS5)

w_C	w_R	T^*	s^*	τ^*	$C(T^*, s^*)$	Utility
0.9	0.1	32.72	17.34	15.39	235.22	−0.90
0.7	0.3	32.72	17.34	15.39	235.22	−0.61
0.5	0.5	32.72	17.34	15.39	235.22	−0.31
0.3	0.7	32.72	17.34	15.39	235.22	−0.02
0.1	0.9	32.72	17.34	15.39	235.22	0.27

5.4.4 Conclusion

We discussed the optimal software release problem based on the MAUT. We defined the cost and reliability attributes based on the change-point models and formulated the weighted multi-attribute utility function as an optimization problem considering multiple evaluation criteria. By solving the optimization problem, we estimated the optimal software release time, the testing duration from the change-point to the testing termination time (optimal change-point), the expected total software cost, and the utility. Furthermore, we reported the result of sensitivity analysis demonstrating the effect of the cost and reliability attributes on the behavior of the total software cost and the utility.

5.5 CONCLUSION

We discussed the change-point modeling and the optimal software release problem in software development management. The obtained main contributions and future studies for the respective sections are summarized as follows.

- Section 5.2 described the proposed change-point models based on the existing exponential, delayed S-shaped, and inflection S-shaped SRGMs. It also proposed several change-point models considering the uncertainty of the testing environment. We demonstrated that these models achieve a better fitting performance compared with the existing models in terms of the MSE. Therefore, we can say that our models describe the fault-occurrence process in the actual testing environment. As a future study, the proposed models need to be compared in terms of the goodness of fit using many actual data sets. In addition, we need to develop various testing-environment functions and check the performance of our models applied to them. Furthermore, we proposed a change-point detection method based on the Laplace trend test. In particular, we estimated the change-points by analyzing the behavior of the Laplace factor and checked the performance of our change-point models using the detected change-points. The result of a numerical example confirmed that the proposed change-point models achieve a better fitting performance. However, the detection method did not demonstrate complete consistency, because the relationship between the change-point and the Laplace factor was not described theoretically. Therefore, we need to correlate them mathematically.

- Section 5.3 discussed the optimal software release problem based on the change-point model. We analytically derived the optimal software release time and the optimal testing duration from the change-point to the testing termination time while minimizing the total software cost. After that, we evaluated them using a reliability-evaluation criterion and determined the conclusive optimal software release time. However, the optimal software release time for a partial data set had not been properly estimated. Therefore, we need to improve the optimal software release policy.

- Section 5.4 discussed the optimal software release problem based on the change-point model and the MAUT. The cost and reliability attributes based on the change-point models were considered as the evaluation criteria, and the optimal software release time and the optimal testing duration from the change-point to the testing termination time were determined simultaneously on the basis of the management strategy. In addition, we derived the optimal change-point, the expected total software cost, and the utility. As a future study, we need to consider a utility evaluation method since our approach does not have a specific evaluation criterion. In other words, it is difficult to evaluate the result by absolute evaluation, because the utility does not have a unit. Therefore, we need to discuss a specific evaluation method applied to software development management.

It is easy for software development managers to apply the proposed techniques to the actual testing environment. Furthermore, the MAUT is a new approach to the optimal software release problem, and it is expected to be discussed more from the point of view of economics. As a future study, we will discuss the aforementioned problems and improve accuracy of the proposed models and the existing software development management methods.

REFERENCES

Chang, Y.-P. "Estimation of Parameters for Non-Homogeneous Poisson Process: Software Reliability with Change-Point Model." *Communications in Statistics: Simulation and Computation* 30(3), 2001: 623–635.

Gaudoin, O. "Optimal Properties of the Laplace Trend Test for Soft-Reliability Models." *IEEE Transactions on Reliability* 41(4), 1992: 525–532.

Gokhale, S. S., and K. S. Trivedi. "Log-Logistic Software Reliability Growth Model." *Proceedings of the Third IEEE International High-Assurance Systems Engineering Symposium*, 1998, pp. 34–41. IEEE, 1998.

Huang, C.-Y. "Performance Analysis of Software Reliability Growth Models with Testing-Effort and Change-Point." *Journal of Systems and Software* 76(2), 2005a: 181–194.

Huang, C.-Y. "Cost-Reliability-Optimal Release Policy for Software Reliability Models Incorporating Improvements in Testing Efficiency." *Journal of Systems and Software* 77(2), 2005b: 139–155.

Huang, C.-Y., and C.-T. Lin. "Analysis of Software Reliability Modeling Considering Testing Compression Factor and Failure-to-Fault Relationship." *IEEE Transactions on Computers* 59(2), 2010: 283–288.

Huang, C.-Y., and T.-Y. Hung. "Software Reliability Analysis and Assessment Using Queueing Models with Multiple Change-Points." *Computers & Mathematics with Applications* 60(7), 2010: 2015–2030.

Inoue, S., and S. Yamada. "Software Reliability Measurement with Change-Point." *Proceedings of International Conference on Quality and Reliability (ICQR 2007)*, pp. 170–175. 2007.

Inoue, S., and S. Yamada. "Optimal Software Release Policy with Change-Point." *IEEE International Conference on Industrial Engineering and Engineering Management*, 2008, pp. 531–535. IEEE, 2008.

Inoue, S., and S. Yamada. "Environmental-Function-Based Change-Point Modeling for Soft-Ware Reliability Measurement." *Proceedings of the Tenth International Conference on Industrial Management*, Beijing, China, September 16–18, 2010, pp. 403–407.

Inoue, S., and S. Yamada. "Software Reliability Growth Modeling Frameworks with Change of Testing-Environment." *International Journal of Reliability, Quality and Safety Engineering* 18(4), 2011: 365–376.

Kapur, P. K., S. K. Khatri, A. Tickoo, and O. Shatnawi. "Release Time Determination Depending on Number of Test Runs Using Multi Attribute Utility Theory." *International Journal of System Assurance Engineering and Management* 5(2), 2014: 186–194.

Kapur, P. K., V. B. Singh, O. Singh, and J. N. P. Singh. "Software Release Time Based on Different Multi-Attribute Utility Functions." *International Journal of Reliability, Quality and Safety Engineering* 20(04), 2013: 1350012.

Langberg, N., and N. D. Singpurwalla. "A Unification of Some Software Reliability Models." *SIAM Journal on Scientific and Statistical Computing* 6(3), 1985: 781–790.

Lin, C.-T., and C.-Y. Huang. "Enhancing and Measuring the Predictive Capabilities of Testing-Effort Dependent Software Reliability Models." *Journal of Systems and Software* 81(6), 2008: 1025–1038.

Lyu, M. R. *Handbook of Software Reliability Engineering*. Vol. 222. IEEE Computer Society Press, CA, 1996.

Minamino, Y., S. Inoue, and S. Yamada. "NHPP Models for Software Reliability Measurement with Change-Point." *Proceedings of the 17th ISSAT International Conference on Reliability and Quality in Design*, Vancouver, Canada, August 4–6, 2011, pp. 132–136.

Minamino, Y., S. Inoue, and S. Yamada. "NHPP Models with Change-Point for Software Reliability Assessment and Its Application to an Optimal Software Release Problem." *Proceedings of the 19th ISSAT International Conference on Reliability and Quality in Design*, Honolulu, Hawaii, August 5–7, 2013, pp. 93–97.

Minamino, Y., S. Inoue, and S. Yamada. "On Application Methodologies of Software Reliability Model with Change-Point." *Asia-Pacific Journal of Industrial Management (APJIM)*, 5(1), 2014a: 63–70.

Minamino, Y., S. Inoue, and S. Yamada. "Change-Point Modeling and Detection Method for Software Reliability Assessment." *Proceedings of the 12th International Conference on Industrial Management (ICIM)*, Chendu, China, September 3–5, 2014b, pp. 266–270.

Minamino, Y., S. Inoue, and S. Yamada. "Multi-Attribute Utility Theory for Estimation of Optimal Release Time and Change-Point." *International Journal of Reliability, Quality and Safety Engineering* 22(4), 2015a: 1550019.

Minamino, Y., S. Inoue, and S. Yamada. "Estimating Optimal Software Release Time Based on a Change-Point Model by Multi-Attribute Utility Theory." *Proceedings of the 21st ISSAT International Conference on Reliability and Quality in Design*, Philadelphia, Pennsylvania, August 6–8, 2015b, pp. 89–93.

Minamino, Y., S. Inoue, and S. Yamada. "NHPP-Based Change-Point Modeling for Software Reliability Assessment and Its Application to Software Development Management." *Annals of Operations Research* 244(1), 2016: 85–101.

Misra, K. B., ed. *Handbook of Performability Engineering*. Springer Science & Business Media, London, 2008.

Okamura, H., T. Dohi, and S. Osaki. "A Reliability Assessment Method for Software Products in Operational Phase—Proposal of an Accelerated Life Testing Model." *Electronics and Communications in Japan (Part III: Fundamental Electronic Science)* 84(8), 2001: 25–33.

Osaki, S. *Stochastic Models in Reliability and Maintenance*. Springer, Berlin, 2002.

Pham, H. *Software Reliability*. Springer Science & Business Media, Singapore, 2000.

Pham, H. *Handbook of Reliability Engineering*. Springer-Verlag, London, 2003.

Pham, H. *Handbook of Engineering Statistics*. Springer-Verlag, London, 2006.

Pham, H., ed. *Recent Advances in Reliability and Quality in Design*. Springer Science & Business Media, London, 2008.

Shyur, H.-J. "A Stochastic Software Reliability Model with Imperfect-Debugging and Change-Point." *Journal of Systems and Software* 66(2), 2003: 135–141.

Singh, O., P. K. Kapur, and A. Anand. "A Multi-Attribute Approach for Release Time and Reliability Trend Analysis of a Software." *International Journal of System Assurance Engineering and Management* 3(3), 2012: 246–254.

Singh, O., P. K. Kapur, A. K. Shrivastava, and V. Kumar. "Release Time Problem with Multiple Constraints." *International Journal of System Assurance Engineering and Management* 6(1), 2015: 83–91.

Teng, X., and H. Pham. "A New Methodology for Predicting Software Reliability in the Random Field Environments." *IEEE Transactions on Reliability* 55(3), 2006: 458–468.

Wang, H., and H. Pham. *Reliability and Optimal Maintenance.* Springer Science & Business Media, London, 2006.

Yamada, S. *Software Reliability Models: Fundamentals and Applications* [in Japanese]. JUSE Press, Tokyo, 1994.

Yamada, S. *Elements of Software Reliability: Modeling Approach* [in Japanese]. Kyoritsu Shuppan, Tokyo, 2011.

Yamada, S. *Software Reliability Modeling: Fundamentals and Applications.* Springer, Tokyo/Heidelberg, 2014.

Yamada, S., and S. Osaki. "Software Reliability Growth Modeling: Models and Applications." *IEEE Transactions on Software Engineering* 12, 1985: 1431–1437.

Yamada, S., and Y. Tamura. *OSS Reliability Measurement and Assessment.* Springer International Publishing, New York, 2016.

Zhao, J., and J. Wang. "Testing the Existence of Change-Point in NHPP Software Reliability Models." *Communications in Statistics—Simulation and Computation* 36(3), 2007: 607–619.

Zhao, M. "Change-Point Problems in Software and Hardware Reliability." *Communications in Statistics: Theory and Methods* 22(3), 1993: 757–768.

CHAPTER 6

An Alternative Approach for Reliability Growth Modeling of a Multi-Upgraded Software System

S. Das, D. Aggrawal, and Adarsh Anand

CONTENTS

6.1 INTRODUCTION

In this competitive environment the role of software has been expanded and impacted every level of society. The growth of the technology sector has advocated users to transform all activities into a digital platform. The applicability of software in the virtually connected economy has increased the core business operation of various organizations day by day. For this very reason the software has to give a consistent performance in every task. Because software is embedded in our daily activities, their failures are even more critical. Reports of tragic consequences of software systems are in ample numbers. As a result, the software development and testing teams have to effectively and efficiently capture all errors in order to maintain the quality and reliability of the software (Kapur et al. 2011a,b). The software testing team tries to detect errors present in software in order to improve the quality of the software. Reliability is one of the foremost important attributes of quality. The reliability of software can be defined as the probability of failure-free operation for a specific period of time under a predefined environment (Kapur et al. 2011a,b). The majority of complex software systems consider reliability as one essential factor to enumerate failures during the development and quality control process.

Specifically, software testing contributes an important role during the software development process. It has to ensure that once the software is up for the release, it has to be free from bugs. The quality of software experienced by the users is a junction of the faults that occur during the operational phase and the efforts made by the software development team to fix it on immediate basis. In order to survive, software developers perform rigorous testing to provide its users a qualitative product. One more significant attribute that causes the adoption of a software product in the highly competitive market is to offer an upgraded version of software. To increase the market presence of the software, firms release new functionalities by periodically providing upgraded versions.

It is well known that a new technology over its lifespan can be described by using the S-shaped or sigmoid curve (Kapur et al. 2011a,b). The initial period requires more effort in order to improve the performance of the technology. After attaining the operational reliability level claimed by the firm, a new version is up for release and the system gets upgraded. The term *software upgrade* means the process of replacement of older versions of software with a newer version of the same product having enhanced features. This phenomenon is most often attributed in computing and electronics products by replacing the software or hardware with a better version to improve the system characteristics. Software upgrading is a complex process. The old and new modules may vary in terms of interface, functionality and performance. But the upgrading process inculcates changes in only selected components while the other parts of the software continue to function. The addition or changes in the software lead to an increase in the fault contents. Therefore, the testing team is always concerned with knowing the flaws present in the software to anticipate adequate resources to inhibit the failure. Although developers schedule upgrades to improve the software application, there is risk involved that may worsen the product after an upgrade. The upgrades may contribute additional bugs and cause the new software to malfunction in some way (Anand et al. 2018).

To capture the software reliability when there is more than one release in the market, various authors have described the modeling framework considering different aspects. Kapur et al. (2010a) considered the impact of faults from all previous releases in the determination fault content in the current release. Along similar lines, Singh et al. (2012) extended the concept given by Kapur et al. (2010a) by limiting the proportion of fault content from all previous releases to just the prior release. Of late, researchers have noticed that the debugging team is not able to remove all faults due to various complexity issues that lead to imperfect fault removal. The concept of imperfect debugging was further analyzed by Kapur et al. (2010b) and Aggrawal et al. (2011) under the frame of multi-upgradation ideology. Furthermore, researchers included some practical aspects in multi-upgradation modeling such as stochastic environment, fault severity and fault removal during the operational phase (Singh et al. 2009, 2011, Kapur et al. 2011a,b, Garmabaki et al. 2014, Anand et al. 2015).

This chapter discusses an important framework to formulate the mean value function for a multi-upgradation software reliability growth model by generalizing a methodical approach to cater the fault from all previous releases in the current offering. Further, the case for four different releases has been discussed. The chapter is organized as follows: In Section 6.2, the methodical approach is presented with mathematical expression for four different releases of the software and the analysis of data is supplemented in Section 6.3. The conclusion is given in Section 6.4.

6.2 MODELING METHODOLOGY

6.2.1 Notation

$m_i(t)$ Mean value function for fault removal phenomena

a_i Total number of faults present in the software ($i = 1, 2, 3, 4$)

b_i Rate of detection/removal of faults ($i = 1, 2, 3, 4$)

β_i Parameter to depict learning of the testing team

Without any loss of generality, we assume that the first release of the software is launched at the time point $t = 0$ and denote the introduction of the ith release as occurring at the end of period $t_i = (i - 1)\tau$. Let $m_i(t)$ denote the number of faults that are being detected/removed in the ith release of the software and is the product of two factors, namely $F_i(t) \times Y_i(t)$, in which $F_i(t)$ denotes the fraction of faults in the ith version of the software that will be debugged in the ith release of the software in period t, and $Y_i(t)$ denotes the number of faults in the ith release of the software consisting of both the faults due to addition of new functionalities and the remaining faults of all previous releases.

First, it is very important to define the exact definition for the fraction of faults:

$$F_i(t) = \begin{cases} 0 & t \leq t_i \\ \dfrac{1 - e^{-b_i(t-t_i)}}{1 + \beta_i . e^{-b_i(t-t_i)}} & t_i < t < t_{i+n} \\ 1 & t > t_{i+n} \end{cases} \qquad (6.1)$$

It is to be noted that n represents the number of releases of a particular software program. If $t < t_i$, then it means that the ith release has not been tested yet, whereas if $t = t_i$, then it means that the software has been given to the testing team to begin its debugging, thus in either case the number bugs discovered is zero. If $t > t_{i+n}$, then the fraction of bugs discovered will be one. Further, if $t_i < t < t_{i+n}$, then there is a positive fraction of faults being debugged for ith release at time t.

Next we consider $Y_i(t)$, which accounts for the number of faults for ith release of the software including the remaining faults of the all preceding releases of the software. Thus we define for $i > 1$.

$$Y_i(t) = Y_{i-1}[1 - F_{i-1}(t_i)] + a_i \qquad (6.2)$$

where the faults for the first release are a_1 (for $i = 1$ we define $Y_1(t) = a_1$) and thereafter the faults of succeeding releases are denoted by a_i, $i > 1$. In other words, $Y_i(t)$ represents remaining faults of all previous releases of the software, $Y_{i-1}(t)F_{i-1}(t)$, plus some incremental growth a_i.

Thus, the actual number of faults debugged in the ith release of the software can be given as follows:

$$m_i(t) = F_i(t)Y_i(t) \qquad (6.3)$$

Equation 6.3 can be interpreted as the fraction $F_i(t)$ of the total faults that are debugged from eventual faults count of any particular release that comprises the faults due to the addition of new functionalities and the remaining faults of all its preceding releases.

Here we have specifically considered the case of software whose four releases were made available to the users. In order to have in-depth understanding, we describe in detail the four releases and the number of faults debugged in each release under the assumption that while debugging faults there are many instances in which remaining faults in previous releases are also removed.

6.2.2 First Release of the Software

Testing is one of the most significant stages in the software development life cycle of any software; it begins as soon as the code has been composed. Firms invest huge amounts of resources to debug the software and to make it bug-free. But in reality it is not possible to make software fault-free. Thus we can say that the testing team may be able to detect a finite proportion of such bugs that it will be able to perfectly remove whose mean value function can be obtained by considering $i = 1$ in Equations 6.1 through 6.3, which yields

$$F_1(t) = \frac{1 - e^{-b_1(t-t_1)}}{1 + \beta_1 . e^{-b_1(t-t_1)}} \quad \text{and} \quad Y_1(t) = a_1$$

Thus, we have the mean value function for fault removal, which is given as follows:

$$m_1(t) = F_1(t)Y_1(t)$$
$$= a_1 . \left[\frac{1 - e^{-b_1(t-t_1)}}{1 + \beta_1 . e^{-b_1(t-t_1)}} \right] \tag{6.4}$$

6.2.3 Second Release of the Software

Due to a technocratic environment, companies are forced to come up with new functionality in the software. These add-ons may lead to complexity and increase in the fault count of the software.

There can be many instances that while debugging the faults of the current release the testing team might be able to come across the latent faults of preceding versions, which eventually leads to increased reliability of the software. Proceeding in a similar manner as the first release of the software in Equations 6.1 through 6.3 we have

$$Y_2(t) = Y_1[1 - F_1(t_2)] + a_2$$
$$= a_1[1 - F_1(t_2)] + a_2 \qquad (6.5)$$

Equation 6.5 accounts for the number of faults in the second release of the software by considering the remaining faults of all previous releases of the software.

$$F_2(t) = \frac{1 - e^{-b_2(t-t_2)}}{1 + \beta_2 . e^{-b_2(t-t_2)}} \qquad (6.6)$$

represents the proportion of faults removed in the second release of the software. Thus, the mean number of faults removed in the testing phase of the second release is given as follows:

$$m_2(t) = F_2(t)Y_2(t)$$
$$= F_2(t)\left[a_2 + a_1.\left(1 - F_1(t_2)\right)\right] \qquad (6.7)$$

where $a_1(1 - F_1(t_2))$ is the amount of leftover faults of the first release that would be debugged using a new fault rate of $F_2(t)$.

6.2.4 Third Release of the Software

Assuming that the firms add new functionalities in the software for the second time this leads to enhanced fault content. These testing phenomena actually enhance the quality of software and also remove some faults of earlier releases, which eventually increase the quality of the software.

Analogous to the modeling for the first and second releases of the software, on considering $i = 3$ in Equation 6.1 through 6.3, we have

$$
\begin{aligned}
Y_3(t) &= Y_2\left[1 - F_2(t_3)\right] + a_3 \\
&= \left[Y_1\left[1 - F_1(t_2)\right] + a_2\right]\left[1 - F_2(t_3)\right] + a_3 \\
&= \left[a_1\left[1 - F_1(t_2)\right] + a_2\right]\left[1 - F_2(t_3)\right] + a_3 \\
&= a_1\left[1 - F_1(t_2)\right]\left[1 - F_2(t_3)\right] + a_2\left[1 - F_2(t_3)\right] + a_3
\end{aligned}
\tag{6.8}
$$

Equation 6.8 accounts for number of faults in the third release of the software when it was assumed that it consists of the remaining faults of all previous releases of the software.

$$
F_3(t) = \frac{1 - e^{-b_3(t - t_3)}}{1 + \beta_3 . e^{-b_3(t - t_3)}}
\tag{6.9}
$$

Equation 6.9 represents the proportion of faults removed in the third release of the software. Thus, the mean number of faults removed in the testing phase of the second release is given as follows:

$$
\begin{aligned}
m_3(t) &= F_3(t)Y_3(t) \\
&= F_3(t)\left[a_3 + a_1\left(1 - F_1(t_2)\right)\left(1 - F_2(t_3)\right) + a_2\left(1 - F_1(t_2)\right)\right]
\end{aligned}
\tag{6.10}
$$

where $a_1(1 - F_1(t_2))(1 - F_2(t_3))$ is the faults leftover from the first and second releases of the software being debugged in the third release with a rate $F_3(t)$.

6.2.5 Fourth Release of the Software

Similar to the case of the third release of the software, when firms add new features for third time in order to attract a large pool of users and for gaining more competitive edge, there are chances that some new faults might get invoked. These newly created faults and some remaining faults of all its preceding release will be removed in testing phase of the fourth release.

Analogous to the modeling for the first and second release of the software, on considering $i = 4$ in Equation 6.1 through 6.3, we have

$$Y_4(t) = Y_3\left[1 - F_3(t_4)\right] + a_4$$
$$= \left[Y_2\left[1 - F_2(t_3)\right] + a_3\right]\left[1 - F_3(t_4)\right] + a_4$$
$$= \left[\left[Y_1\left[1 - F_1(t_2)\right] + a_2\right]\left[1 - F_2(t_3)\right] + a_3\right]\left[1 - F_3(t_4)\right] + a_4$$
$$= \left[\left[a_1\left[1 - F_1(t_2)\right] + a_2\right]\left[1 - F_2(t_3)\right] + a_3\right]\left[1 - F_3(t_4)\right] + a_4$$
$$= \left[a_1\left[1 - F_1(t_2)\right]\left[1 - F_2(t_3)\right] + a_2\left[1 - F_2(t_3)\right] + a_3\right]\left[1 - F_3(t_4)\right] + a_4$$
$$Y_4(t) = a_1\left[1 - F_1(t_2)\right]\left[1 - F_2(t_3)\right]\left[1 - F_3(t_4)\right]$$
$$+ a_2\left[1 - F_2(t_3)\right]\left[1 - F_3(t)\right] + a_3\left[1 - F_3(t_4)\right] + a_4$$

$$(6.11)$$

Equation 6.11 accounts for the number of faults in the third release of the software in consideration of the remaining faults of all previous releases of the software.

$$F_4(t) = \frac{1 - e^{-b_4(t - t_4)}}{1 + \beta_4 . e^{-b_4(t - t_4)}} \qquad (6.12)$$

Equation 6.12 represents the proportion of faults removed in the fourth release of the software. Thus, the mean number of faults removed in the testing phase of fourth release is given as follows:

$$m_4(t) = F_4(t)Y_4(t)$$
$$= F_4(t)\begin{bmatrix} a_4 + a_1\left(1 - F_1(t_2)\right)\left(1 - F_2(t_3)\right)\left(1 - F_3(t_4)\right) \\ + a_2\left(1 - F_1(t_2)\right)\left(1 - F_2(t_3)\right) + a_3\left(1 - F_3(t_4)\right) \end{bmatrix}$$

$$(6.13)$$

where $a_1(1 - F_1(t_2))(1 - F_2(t_3))(1 - F_3(t_4))$ is the remaining faults from the initial software release that would be debugged in the latest release.

6.3 DATA ANALYSIS

The proposed model has been analyzed on tandem software data that consists of four releases (Wood 1996). Parameters have been estimated using the SAS software package (SAS 2004). The values of parameters are represented in Table 6.1 and comparison criteria, that is SSE and MSE, and have shown in Table 6.2. The goodness-of-fit curve shows how the observed data deviated from actual data, which are given in Figures 6.1 through 6.4.

TABLE 6.1   Parameter Estimates of Four Different Releases ($i = 1, 2, 3, 4$)

Parameters	Release 1	Release 2	Release 3	Release 4
a_i	110.205	125.548	68.994	44.999
b_i	0.175	0.256	0.440	0.264
β_i	1.201	3.501	11.876	5.353

TABLE 6.2   Comparison Criteria for Four Releases

Criterion	Release 1	Release 2	Release 3	Release 4
SSE	548.4	288.9	50.7965	17.8561
MSE	3.4642	2.992	1.6441	1.016
R^2	0.989	0.995	0.993	0.995

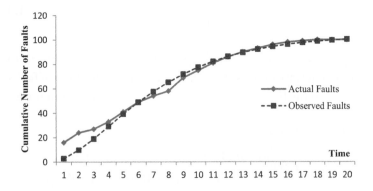

FIGURE 6.1   Goodness-of-fit curve for Release 1.

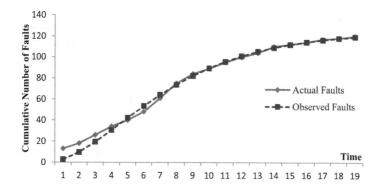

FIGURE 6.2   Goodness-of-fit curve for Release 2.

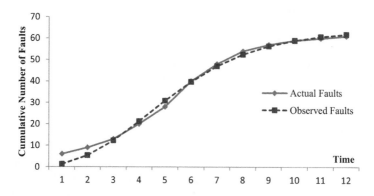

FIGURE 6.3   Goodness-of-fit curve for Release 3.

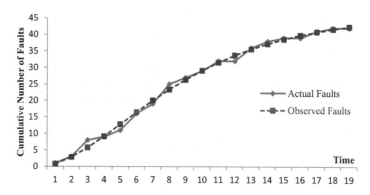

FIGURE 6.4   Goodness-of-fit curve for Release 4.

6.4 CONCLUSION

Today's software firms are interested in capturing the market share through upgradation leading to many more chances of increased fault content in the software wherein it is of utmost importance to make it fault-free. Firms generally keep testing their software to increase its reliability. In this chapter, we proposed a generalized framework that captures the fault count of preceding all releases. The obtained results depict that the proposed models are able to cater to the fault removal phenomenon in a remarkable manner. In the future we can extend the methodology to capture more realistic scenarios, such as imperfect debugging, fault severity, and randomness, to yield a generalized framework.

ACKNOWLEDGMENT

The work illustrated in this article is supported by grants to the first author via Rajiv Gandhi National Fellowship from the University Grants Commission, New Delhi, India.

REFERENCES

Aggarwal, A. G., P. K. Kapur, and A. S. Garmabaki. "Imperfect Debugging Software Reliability Growth Model For Multiple Releases." *Proceedings of the 5th National Conference on Computing for Nation Development—INDIAcom*, New Delhi, India, pp. 337–344. 2011.

Anand, A., P. Gupta, Y. Klochkov, and V. S. S. Yadavalli. "Modeling Software Fault Removal and Vulnerability Detection and Related Patch Release Policy." In: *System Reliability Management*, pp. 35–50. CRC Press, 2018.

Anand, A., O. Singh, and S. Das. "Fault Severity Based Multi Up-Gradation Modeling Considering Testing and Operational Profile." *International Journal of Computer Applications* 124(4), 2015: 9–15.

Garmabaki, A. H. S., P. K. Kapur, A. G. Aggarwal, and V. S. S. Yadavali. "The Impact of Bugs Reported from Operational Phase on Successive Software Releases." *International Journal of Productivity and Quality Management* 14(4), 2014: 423–440.

Kapur, P. K., A. Adarsh, O. Singh, and M. N. Hoda. "Modeling Successive Software Up-Gradations with Faults of Different Severity." *Proceedings of the 5th National Conference on Computing for Nation Development—INDIAcom*, New Delhi, India, pp. 351–356. 2011a.

Kapur, P. K., H. Pham, A. Gupta, and P. C. Jha. *Software Reliability Assessment with OR Applications*. London: Springer, 2011b.

Kapur, P. K., A. Tandon, and G. Kaur. "Multi Up-Gradation Software Reliability Model." *2010 2nd International Conference on Reliability, Safety and Hazard (ICRESH)*, pp. 468–474. IEEE, 2010a.

Kapur, P. K., O. Singh, A. S. Garmabaki, and J. Singh. "Multi up-gradation software reliability growth model with imperfect debugging." *International Journal of System Assurance Engineering and Management* 1(4), 2010b: 299–306.

SAS Institute Inc. *SAS/ETS User's Guide Version 9.1*. Cary, NC: SAS Institute Inc., 2004.

Singh, O., P. K. Kapur, and A. Anand. "A Stochastic Formulation Of Successive Software Releases with Faults Severity." *2011 IEEE International Conference on Industrial Engineering and Engineering Management (IEEM)*, pp. 136–140. IEEE, 2011.

Singh, O., P. K. Kapur, A. Anand, and J. Singh. "Stochastic Differential Equation Based Modeling for Multiple Generations of Software." *Proceedings of Fourth International Conference on Quality, Reliability and Infocom Technology (ICQRIT), Trends and Future Directions*, pp. 122–131. Narosa Publications, 2009.

Singh, O., P. K. Kapur, S. K. Khatri, and J. N. P. Singh. "Software Reliability Growth Modeling for Successive Releases. proceeding of 4th International Conference on Quality." *Reliability and Infocom Technology (ICQRIT)*, pp. 77–87. 2012.

Wood, A. "Predicting Software Reliability." *Computer* 29(11), 1996: 69–77.

CHAPTER 7

Assessing Software Reliability Enhancement Achievable through Testing

Y. K. Malaiya

CONTENTS

7.1 INTRODUCTION

For achieving high reliability in software, we must consider the nature of actual defects and the defect finding process, which is necessary to make testing efficient and to ensure that the reliability growth modeling is realistic. This chapter presents a perspective based on the nature of actual defects as described by the detectability profile, introduced in the next section. The real defects vary significantly in testability and the hard-to-test faults are likely to be found later during testing. The next section considers software partitioning and how it impacts testing effectiveness. The defect-finding process is initially modeled using the common assumptions of a fixed fault exposure ratio and then refined using actual reliability growth data. The two approaches result in the exponential and the logarithmic Poisson software reliability growth models (SRGMs) as discussed in following section. The final section considers the relationship between test coverage and defect density. The last section considers the extent to which ultra-high reliability is achievable.

7.2 THE NATURE OF SOFTWARE DEFECTS

The defects in software arise during different phases of software development. The development process can be regarded as a multistep translation of the software requirements into executable code. In terms of the waterfall model, the steps can be described as (Boehm 1988)

1. Business case

2. Requirements

3. Design

4. Development, including the development of units and unit testing

5. Integration and integration testing

6. System testing including debugging and acceptance testing

7. Deployment

In actual practice, the process may be more complex. There may be a partial release to allow external beta testing. Most successful software products undergo periodic revisions for bug fixing and the addition of features/capabilities. This may be termed the maintenance phase, which includes regression testing that ensures that the existing functionality is not impaired by the revision. With the emergence of Internet-based automated patch delivery, the software can evolve nearly continuously using an Agile (Rawat et al. 2017) or DevOps development approach.

Defects arise because of imperfect translation to a lower level (in terms of the phases mentioned earlier 0–1, 1–2, 2–3 or 3–4). Bugs introduced during the 0-to-1 translation can arise not only because of the imperfect capture of the business case into requirements, but also shifts in the business case. Modern Agile or DevOps approaches attempt to minimize their impact. The 0-to-1 and 1-to-2 translations are higher level and can give rise to bugs that are harder to identify and expensive to fix. Available data suggests that most bugs occur at phases 2 and 3. Some defects may occur during integration because the module functionality or the interface requirements may not be clearly understood.

A bug that involves only a single line of code, or a few adjacent lines, are easier to find and debug. Some bugs can arise because of imperfect implementation of a computational task that involves parts of the code that are not adjacent. Such faults may be triggered only under some specific circumstances ("corner cases") and may not be easily detected.

A defect (also termed a bug) is defined as a deviation of a set of statements (not necessarily contiguous) from the higher level need that may be reported as a single *trouble report* and need to be fixed as a single corrective action. We will initially assume that when a defect is encountered during testing, the associated code is fixed ("debugged"). During the actual operation, failures may be recorded (to be addressed by the next patch or release) but are not fixed immediately. Thus during testing, reliability growth occurs; during normal operation, the reliability does not change.

When a code segment containing a defect is exercised, an *error* may be generated. The error may cause incorrect data or an improper execution sequence. When the error propagates to the output, a *failure* is said to have occurred.

There are a few measures that are often used to describe software reliability. For any non-trivial real-life software, it is virtually impossible to ensure that the software is bug-free. It may possible to use formal methods for small and well-defined pieces of code. That requires the software to be described with mathematical rigor and then mathematically proven to be correct. For most practical software development that is not feasible. The common software reliability measures used are

- *Failure rate*: The rate at which the failures are encountered. In the formal software reliability literature, it is represented by the failure intensity.

- *Defect density*: The number of defects per thousand lines of code (without counting the comments).

- *Transaction reliability*: It is the probability that a transaction, which takes a limited amount of execution time, is executed correctly without a failure.

These measures are used in the following discussion.

7.3 DETECTABILITY PROFILE

The effectiveness of software testing techniques and the failure rate encountered during normal operation depend on the nature of the defects in the software. Some bugs can be very easy to find. For example, some bugs that violate the syntax requirements may be detected during compilation; they may not even arise when a continuously compiling development environment (such as Eclipse) is used. On the other hand, some bugs may affect computation only when a specific combination of rarely occurring inputs is encountered. The detectability of a defect may be measured by how hard it is to test. The detectability of a bug must be defined in terms of random testing (Malaiya and Yang 1984). An input combination is random if each input value is chosen randomly and independently. In general, the bugs in a program will have different detectability values. The distribution of the detectability values in the program will determine how hard it will be to debug the code to arrive at a target defect density.

If there the total number of distinct combinations is N and j out of them detect the fault f_i then f_i has detectability j/N, since

Pr{defect f_i is detected by a randomly applied input combination} $= N/j$

Thus we can write

$$
\Pr\left(\begin{array}{l} n \text{ randomly applied input combinations} \\ \text{do not detect the fault } f_i \end{array}\right)
$$

$$
= \left(1 - \frac{j}{N}\right)^n \tag{7.1}
$$

Note that multiple faults may have the same detectability value. The discrete *detectability profile* is a vector $\{\pi_1, \pi_2, \pi_3, \ldots, \pi_N\}$ that gives the number for faults with detectability values $\{1/N, 2/N, 3/N, \ldots, N/N\}$. The faults in the set π_1 are hardest to test, since they are each detectable only by a single specific test. Let the total number of faults be $M = \sum_{i=1}^{N} \pi_i$. Faults in the set π_N are the easiest to test since any randomly chosen test will test for the faults. It can be shown the expected fraction of faults that would be covered by L random tests would be given by (Malaiya and Yang 1984)

$$C(L) = 1 - \sum_{k=1}^{N}\left(1 - \frac{k}{N}\right)^{L}\frac{\pi_k}{M} \qquad (7.2)$$

If the tests applied are pseudorandom (when tests are chosen without replacement), it can be shown (Wagner et al. 1987) that the expected coverage is

$$C(L) = 1 - \sum_{k=1}^{N-L}\frac{\binom{N-L}{k}}{\binom{N}{k}}\frac{\pi_k}{M} \qquad (7.3)$$

As software undergoes testing and removal of bugs, the bugs with a high detectability will be encountered earlier and removed. That will leave bugs with low detectability behind. Thus as testing progresses, the detectability profile of the software will continue to shift. The two preceding expressions show that once the test coverage is sufficiently high, only the defects with low values of (k/N) would matter. Sometimes for convenience, the detectability profile may be defined continuously in terms of the test time instead of the number of tests.

At practically any point in time during the development, easy-to-find bugs would have been removed because of some prior testing. During the last phases of testing, the remaining faults are likely to be those that are triggered only by rarely occurring combinations of inputs. The detectability profile would thus shift and become increasingly asymmetric. An example can be seen in the data published by Adams (1984) for a large IBM project, as seen in Figure 7.1.

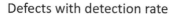

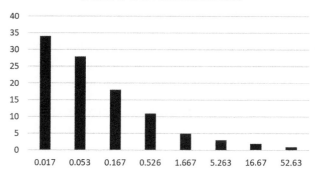

FIGURE 7.1 Testability profile for Adam's data.

7.4 SOFTWARE PARTITIONING

To ensure that the software is thoroughly exercised during testing, it is generally necessary to partition it to identify tests that would be effective for detecting the defects in different sections of the code. For testing purposes, a program may be partitioned either functionally or structurally.

- *Functional partitioning* refers to partitioning the input space of a program. For example, if a program performs five separate operations, its input space can be partitioned into five partitions. Functional partitioning only requires the knowledge of the functional description of the program; the actual implementation of the code is not required.

- *Structural partitioning* requires the knowledge of the structure at the code level. If a software program is composed of 10 modules (which may be classes, functions or other types of units), it can be thought of as having 10 partitions.

A partition of either type can be subdivided into lower-level partitions, which may themselves be further partitionable at a lower level if higher resolution is needed (Elbaum and Narla 2001). Dividing a partition into a lower partition has the following consequences. Let us assume that a partition p_i can be subdivided into sub-partitions $\{p_{i1}, p_{i2} \ldots p_{in}\}$.

- Random testing within the partition p_i will randomly select from $\{p_{i1}, p_{i2} \ldots p_{in}\}$. It is possible that some of them will get selected more often in a nonoptimal manner.

- Code within a sub-partition may be correlated relative to the probability of exercising some faults. Thus the effectiveness of testing may be diluted if the same sub-partition frequently gets chosen.

- Sub-partitioning has a practical disadvantage that when the operational profile is constructed, it will require estimating the operational probabilities of the associated sub-partitions.

For structural partitions, a statement or a branch (which are attributes often measured using software test coverage tools like JCov or Emma) may be regarded as a low-level partition. It should be noted that the execution of the statements within a straight-line *block* (containing statements that are always executed one after the other) is completely correlated. Thu if a single block is partitioned into multiple partitions, their execution and potential discovery of related defects will be completely correlated.

An *operational profile*, as defined by John Musa (1993), involves the use of functional partitioning. An operation profile is a set of functional partitions along with a probability associated with each partition that gives the probability that an input is drawn from that partition. More elaborate operational profiles can be constructed by considering the system states using a state diagram and the transition probabilities (Regnell et al. 2000). The actual resolution used in the operational profile can vary significantly. Guen and Thelin (2003) report the partitions to range from 4 to 143 partitions per thousand lines of code. Operation profile–based testing is also sometimes termed statistical usage testing (Runeson and Wohlin 1995).

TABLE 7.1   Sub-Partitioning a Partition

Initial		Refined	
Operation	**Probability**	**Operation**	**Probability**
P1: Voice call	0.74	Voice call, no pager, answer	0.18
		Voice call, no pager, no answer	0.17
		Voice call, pager, voice answer	0.17
		Voice call, pager, answer on page	0.12
		Voice call, pager, no answer on page	0.10
FAX call	0.15	FAX call	0.15
New number entry	0.10	New number entry	0.10
Data base audit	0.009	Data base audit	0.009
Add subscriber	0.0005	Add subscriber	0.0005
Delete subscriber	0.000499	Delete subscriber	0.000499
Failure recovery	0.000001	Failure recovery	0.000001

A simple example of incremental refinement is provided by Musa. A PABX unit has the initial operational profile given in column 1 of Table 7.1. Since partition P1 has a large probability associated with it, it can be divided into sub-partitions as shown in column Refined.

When a software project is still in the unit development/ testing phase and has not been integrated, each module can be considered both a functional partition and a structural partition. After integration has taken place, some of the code (and associated defects) may be shared by the functional partitions. If the shared code is small and has been thoroughly tested and debugged, then the bugs associated with each functional partition may be essentially disjoint, since they are in disjoint structural partitions.

Test and operation profiles. During testing, the testing profile may or may not correspond to the operational profile. Using the operational profile is preferred in these two cases:

1. Acceptance testing, to assess the failure intensity that would be encountered during actual operation.

2. When the testing time available is limited. In this case, the impact of testing would be maximized by drawing more inputs from a partition that is encountered more during actual operation.

Operational profile–based testing will not be effective in the following cases.

1. Once most bugs from the frequently executed partitions have been removed, partitions that are exercised less often will tend to have a higher defect density. Testing will become inefficient if these partitions are still exercised less frequently (Li and Malaiya 1994).

2. Some of the partitions may represent reused code, which has already undergone prior testing during the testing for prior releases. In this case, testing should focus on new code (Malaiya 2011, 2018).

3. Some partitions may represent critical operations such that their failure may have a high impact.

4. If a partition corresponds to a larger code segment, it will require testing for a longer time to achieve the factor of reduction in defect density.

EXAMPLE 7.1

This illustrates the case when there are five partitions $P1$ to $P5$. In Table 7.2, the size of each partition (measured in KLOC),

TABLE 7.2   Table for Example 7.1 with the Five Partitions

Partition/Attribute	P1	P2	P3	P4	P5
Size in KLOC	1	5	3	1	5
DefDensity	5	5	10	20	20
ExecutionFreq	0.1	0.3	0.2	0.1	0.3

the initial defect density and the execution frequencies are given. We assume that the partitions are disjoint. Note that the operational profile is {$P1$, $P2$, $P3$, $P4$, $P5$} = {0.1, 0.3, 0.2, 0.1, 0.3}. Thus 30% of the time an input from $P2$ is chosen. The total size of the code 15 KLOC. We use these numbers for later examples.

Musa, Iannino and Okumoto have defined the testing compression factor (TCF) as the ratio of the execution times needed to cover all the partitions (they use the term *state*) during testing and during normal operation. This ratio can be used as a measure of the effectiveness of testing compared with operational use (Huang and Lin 2010).

$$\text{TCF} = \frac{\text{Partitions exercised per unit time during testing}}{\text{Partitions exercised per unit time during operation}}$$

(7.4)

Musa found that the value for the TCF is between 8 and 21 for many of the programs considered. In a situation where the defects are uniformly distributed among the partitions, the defect finding rate, and hence the failure intensity during testing, would be accelerated by a factor of TCF compared with operational use.

7.5 RELIABILITY GROWTH MODELS AND THE SIGNIFICANCE OF MODEL PARAMETERS

Here we consider the significance of the parameters of the common exponential model and see how they can be used for optimal test effort distribution for a software. We then examine the variation in the fault exposure ratio and show that the variation results in the logarithmic Poisson model, which has been shown to provide better predictability.

Researchers have proposed a number of software reliability growth models. The simplest of them can be termed the exponential model, which can be thought to represent several models proposed earlier including Jelinski and Muranda (in 1971), Shooman (in 1971), Goel and Okumoto (in 1979) and Musa (between 1975 and 1980) (Musa et al. 1987, Yamada 2014). It has the advantage of offering straightforward interpretations of the model parameters in terms of measurable physical quantities.

The exponential model is based on the assumption that the defect finding rate at any point is proportional to the number of defects remaining at that time. Let us denote the number of yet undetected defects at time t to be $N(t)$. The testing effort is measured in terms of time t. It can be the CPU time (as used by Musa), calendar time, or some other measure such as operational coverage (Kansal et al. 2018) or the testing effort function (Peng et al. 2014).

Initially, we assume that debugging is perfect, implying that a defect is always successfully removed when it is encountered.

Let $N(t)$ be the number of defects that has remained undetected at time t. Let T_s be the execution time of a test case, and let k_s be the fraction of faults found during a single test case. Then using the aforementioned assumption, we can write

$$-\frac{dN(t)}{dt}T_s = k_s N(t) \tag{7.5}$$

For the convenience of notation, let us introduce a term K called *fault exposure ratio*, such that

$$K = k_s \frac{T_L}{T_s} \tag{7.6}$$

where $T_L = S \cdot Q \cdot (1/r)$, where S is the source code size, Q is the number of object instructions per source instruction, and r is the instruction execution rate of the computer. Here we assume that the testing time t is measured in terms of the CPU execution time. The fault exposure ratio is a metric that measures the fault exposing capabilities of the testing strategy. Musa has argued that it should be independent of the software size S. The exponential model assumes that the fault exposure ratio remains constant throughout testing. We examine the assumption later.

Using Equation 7.6, Equation 7.5 can be written as

$$-\frac{dN(t)}{dt} = \frac{K}{T_L}N(t) \tag{7.7}$$

Following Musa's notation, let us indicate the ratio (K/T_L) by β_1, which will serve as one of the model parameters. Solving the differential equation gives us

$$N(t) = N(0)e^{-\beta_1 t} \tag{7.8}$$

The failure intensity, which describes the defect finding rate, is given by

$$\lambda(t) = -\frac{dN(t)}{dt}$$

Thus,

$$\lambda(t) = \beta_0 \beta_1 e^{-\beta_1 t} \tag{7.9}$$

where the initial number of defects $N(0)$ serves as the other parameter β_0. The mean value function $\mu(t)$ expresses the cumulative expected number of defects found by time t and is thus given by

$$\mu(t) = \beta_0(1 - e^{-\beta_1 t}) \tag{7.10}$$

Equations 7.9 and 7.10 express the two forms of the exponential model. The two parameters can be readily interpreted (Malaiya and Denton 1997).

Parameter β_0. In Equation 7.9, $\mu(t)$ approaches β_0 as t approaches infinity. Thus it is the total number of defects that would eventually be detected. If no defects were injected during debugging, it will be equal to $N(0)$. In actual practice debugging is imperfect and some bugs are injected during debugging. Studies have shown that the number of such injected defects can be in the range of 5%, causing the final value of β_0 to be somewhat higher. In many projects, the defect density can be estimated using previous projects and perhaps using some of the static metrics. If the defect density at the onset of testing is $D(0)$, then

$$\beta_0 = D(0) \cdot S \tag{7.11}$$

Parameter β_1. Equation 7.6 provides an interpretation of the parameter β_1, which can be written as

$$\beta_1 = \frac{K}{T_L} = \frac{K}{S \cdot Q \cdot \dfrac{1}{r}} \tag{7.12}$$

The data sets collected by Musa suggest the values of K ranging between 1×10^{-7} and 10×10^{-7}. When the data is collected using some other measure of the testing effort (such as person-hours, etc.), the value of β_1 should be multiplied by an appropriate factor. Note that Q depends on the high level language and machine architecture, and r is machine dependent. Thus β_1 is proportional to the testing efficiency and inversely proportional to the software size S. Also notable is the fact that since the initial failure intensity is the product $\beta_0\beta_1$, it would be independent of the software size.

EXAMPLE 7.2

For the five partitions $P1$ to $P5$, we assume that the fault exposure ratio is 5×10^{-7}, the object instructions per source instruction is 2.5 and the object instruction execution rate for the processor is 7×10^7 per second. Then from

TABLE 7.3   Table for Example 7.2 Illustrating the Computation of the
Parameters' Values

Partition/Attribute	P1	P2	P3	P4	P5
Size in KLOC	1	5	3	1	5
Def Density	5	5	10	20	20
Execution Freq	0.1	0.3	0.2	0.1	0.3
β_0	5	25	30	20	100
β_1		0.014	0.0028	0.004667	0.014	0.0028

Note: β_0 is simply the number of defects in each partition and β_1 depends
inversely on software size.

Equations 7.11 and 7.12, we can estimate the two parameters
for the exponential model as shown in Table 7.3.

Test time required. The *when to stop testing problem* requires
obtaining the answer to the following question: How much
testing is needed to bring the failure intensity (or equivalently, the
defect density) down to the acceptable threshold? Normalizing
Equation 7.8 by dividing both sides by the software size S, we get

$$D(t) = D_0 e^{-\beta_1 t} \tag{7.13}$$

where $D_0 = D(0)$ is the initial defect density. If the target defect
density is D_T, then we can obtain the test time needed as

$$t_F = \frac{-\ln\left(\dfrac{D_T}{D_0}\right)}{\beta_1} \tag{7.14}$$

Equation 7.14 implies that more testing time is needed to reach
the target if the initial defect density is higher. Also, since β_1 is
inversely proportional to size, a larger module will need to be
exercised for a longer time.

Optimal testing. Testing using the operational profile is not
always the most effective approach for debugging and thus
achieving higher reliability. This is illustrated here in the next

example. Here we assume that the overall failure rate is the weighted sum of the individual failure rates:

$$\lambda_{sys} = \sum_{i=1}^{n} f_i \lambda_i \qquad (7.15)$$

where λ_i is the failure rate for partition i and f_i is the fraction of time i is under execution.

EXAMPLE 7.3

This example uses the parameter values as estimated in Example 7.2. We can set this up as an optimization problem (Malaiya 2018) with the overall failure rate as the objective function, and maximum allowable testing time (chosen to be 1500 units here) as a constraint. The problem is to allocate the 1500 units of testing to the five partitions. The optimal results can be obtained using the algebraic Lagrange multiplier technique, which yields a closed-form solution, or using an iterative algorithm (for example, using the Microsoft Excel Solver).

Table 7.4 shows testing times allocated and gives the resulting system failure rate if operational profile testing is done, and when optimization is done.

TABLE 7.4   Table for Example 7.3: Operational Profile Based versus Optimal Testing

Partition/Attribute	P1	P2	P3	P4	P5	λ_{sys}
β_0	5	25	30	20	100	
β_1	0.014	0.0028	0.004667	0.014	0.0028	
			Op Profile Testing			
Testing time	150	450	300	150	450	
Failure rates	0.0086	0.0199	0.0345	0.0343	0.0794	**0.0410**
			Optimal Testing			
Testing time	80.6023	220.5986	303.4676	179.6356	715.6959	
Failure rates	0.0226	0.0377	0.0340	0.0226	0.0377	**0.03397**

It can be seen that the optimal distribution of the test effort is significantly different from the operational profile based testing. Part of this is due to the fact the *P1* and *P2* have lower defect densities, and *P4* and *P5* have higher defect densities. The code size in each partition also makes a significant impact.

Variation of the fault exposure ratio K. The preceding discussion uses the simplifying assumption that the fault exposure ratio is constant throughout testing. The assumption can be questioned because of these two facts:

1. As testing progresses, faults that are easy to find are found and removed, leaving faults with lower detectability. This will cause detection efficiency to decline.

2. In truly random testing, each test case applied is chosen regardless of the previous tests applied. In actual practice, the test scheme may remember the partitions that have been exercised in the past and focus on partitions not yet exercised. This will cause the testing efficiency to go up when fewer unexercised partitions remain and testing focuses on them.

Quantitative examination of the test data from several projects suggests that the fault exposure ratio does vary (Malaiya et al. 1993). It declines when the initial defect density is high, and eventually starts rising when testing has progressed sufficiently far.

Malaiya et al. have examined the variation of the fault exposure ratio K for 13 industrial reliability growth data sets. They observed that at higher fault densities K declines, whereas at lower fault densities K tends to rise as testing progresses. The change appears to occur in the vicinity of density about 2 per KLOC, although it is likely to vary depending on the testing approach used.

Considering the fact that the faults remaining undetected tend to be the ones harder to find, it can be shown (Malaiya et al. 1993) that the variation in K can be approximately modeled by

$$K(t) = \frac{K(0)}{1+at}, \quad \text{where } a > 0 \tag{7.16}$$

where a is a parameter. The impact of testing becoming more focused on the partitions not yet covered can be assessed by considering the extreme case when the location of the faults is known, assuming that the application of each test has the same likelihood of revealing the presence of a new fault. In this situation, we have

$$\frac{dN}{dt} = -C$$

where C is a parameter. Based on Equation 7.7, we can obtain

$$K = T_L C \cdot \frac{1}{N} \tag{7.17}$$

Considering both effects, we can hypothesize that the overall variation of $K(t)$ can be represented using a combination of the two factors:

$$K(t) = \frac{g}{N(t)(1+at)} \tag{7.18}$$

where g is a parameter that can be evaluated using $K(0)N(0)$. Substituting this expression for $K(t)$ in Equation 7.7, and solving the differential equation, we obtain

$$N = N(0) - \frac{g}{T_L} \ln(1+at) \tag{7.19}$$

which confirms with the logarithmic Poisson model, which has been found to provide better predictability than the exponential model (Malaiya et al. 1992). The correspondence provides an interpretation for the two parameters of the logarithmic Poisson model (for consistency with the literature and for convenience, we are designating the two parameters using the same notation, even though they are different from the exponential model parameters). They are given by

$$\beta_0 = \frac{g}{T_L} = \frac{K(0)N(0)}{T_L} \tag{7.20}$$

$$\beta_1 = a \tag{7.21}$$

Table 7.5 gives the overall values of K (in units of 10^{-7}) for nine data sets collected by Musa, arranged in the order of decreasing defect densities (in defects per KLOC) (Malaiya et al. 1993). The size is given in KLOC.

Table 7.5 and the plot in Figure 7.2 illustrate the observation that the fault exposure ration initially declines as faults get harder to find and then starts rising due to the use of directed testing in actual industrial testing.

TABLE 7.5   Values of K (Units of 10^{-7}) for the Nine Data Sets

Data Set	Size	D_0	K
T1	21.7	6.89	1.87
T2	27.7	2.14	2.15
T3	23.4	1.79	4.11
T4	33.5	1.74	10.6
T5	2445	0.374	4.2
T6	5.7	14.08	3.97
T16	126.1	0.357	3.03
T19	61.9	0.675	4.54
T20	115.35	20.89	6.5

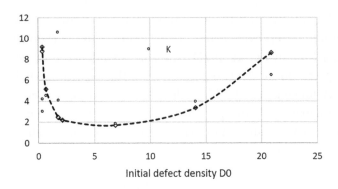

FIGURE 7.2   Variation of K with defect density.

Equation 7.18 gives K in terms of testing time. It can be argued that it should be a function of the defect density, where it declines at higher defect densities and later starts rising at lower defect densities. An expression for $K(D)$ can be obtained as (Li and Malaiya 1996)

$$K = \frac{\alpha_0}{D} e^{\alpha_1 D} \qquad (7.22)$$

where α_0 and α_1 are applicable parameters. It can be shown that the equation also applies for initial defect density D_0 when the model is applied to multiple data sets. The plot in Figure 7.2 gives the actual data points as well as fitted data points.

7.6 COVERAGE-BASED MODELING FOR DEFECT DISCOVERY

During testing, the strategy often changes. It will give rise to bursts in failure intensity. A new strategy may exercise some parts of the code that have not been exercised before. Thus the efficiency of the testing strategy can vary. The SRGMs assume that the testing strategy remains unchanged and uses time as a variable determining the reliability growth. It can be argued that test coverage is a better metric than time since it directly measures the number of test elements exercised. The statements and branches are among the lowest levels of structural partitions. One measure of test effectiveness can be the coverage of the fraction of statements and branches. Two of the most common coverage measures are (Horgan and Mathur 1996)

- *Statement (or block) coverage*: The fraction of the total number of statements (blocks) that have been executed by the test data. A block is a segment of the code in which the instructions are always executed together.

- *Branch (or decision) coverage*: The fraction of the total number of branches that have been executed by the test data.

Weyuker (1993) has shown that the branch coverage subsumes the block coverage, that is, if all the branches have been exercised that guarantees that all the blocks would also have been exercised but not vice versa.

We can obtain a model describing the relationship of the defect coverage with a test coverage metric by combining (1) a model relating defects found and the test time, and (2) a model relating the coverage achieved and the test time. For the first one, we assume that the reliability growth is given by the logarithmic Poisson model. For the second one also we assume that the coverage growth is also modeled by a logarithmic Poisson model. For convenience we use superscript 0 to indicate defects covered, and superscripts 1 and 2 for the statement and branch coverage. (Malaiya et al. 2002). The test or defect coverage is given by

$$C^i(t) = \frac{1}{N^i} \beta_0^i \ln\left(1 + \beta_1^i t\right), \quad C^i(t) \le 1 \qquad (7.23)$$

Note that the logarithmic Poisson is applicable only until all the defects (or statements or branches) have been covered. Here N^i is the total number of enumerables (defects, statements or branches) of type i, and β_0^i and β_1^i are the model parameters. If a single test takes T_s seconds, then the time needed to apply n tests is $T_s n$. Then Equation 7.23 can be written as

$$C^i(n) = \frac{\beta_0^i}{N^i} \ln\left(1 + \beta_1^i T_s n\right) \tag{7.24}$$

Note that for defect coverage, the parameter values are given by

$$\beta_0^0 = \frac{K^0(0)N^0(0)}{a^0 T_L} \tag{7.25}$$

$$\beta_1^0 = a^0 \tag{7.26}$$

For a compact notation, let us denote $\left(\beta_0^i/N^i\right)$ and $\beta_1^i T_s$ by b_0^i and b_1^i, respectively, allowing us to write the preceding equation as

$$C^i(n) = b_0^i \ln\left(1 + b_1^i n\right), \quad C^i(n) \le 1 \tag{7.27}$$

Here we can eliminate the number of vectors n in the expression for $C^0(N)$ by using the expression for test coverage $C^i(n)$, $i = 1, 2$. We get

$$C^0 = b_0^0 \ln\left[1 + \frac{b_1^0}{b_1^i}\left(exp\left(\frac{C^i}{b_0^i}\right) - 1\right)\right], i = 1...2 \tag{7.28}$$

Again for convenience, we can denote $b_0^0, \dfrac{b_1^0}{b_1^0}$ and $\dfrac{1}{b_0^i}$ by parameters a_0^i, a_1^i and a_2^i, respectively, to write

$$C^0 = a_0^i \ln\left[1 + a_1^i\left(exp\left(a_2^i C^i\right) - 1\right)\right], i = 1, 2 \tag{7.29}$$

Equation 7.29 gives an expression for defect coverage in terms of the test coverage. It can be seen that if test coverage is closer to 1, the preceding equation can be approximated by (Malaiya and Denton 1998, Malaiya et al. 2002)

$$C^0 = a_0^i \ln\left(a_1^i\right) + a_0^i a_2^i C^i, \, i = 1,2, C^i > C_{knee}^i \qquad (7.30)$$

where C_{knee}^i is the test coverage level at which the linear trend begins. The plot in Figure 7.3 demonstrates the model given in Equations 7.29 and 7.30. The data is from a European Space Agency project with 6100 lines of C code. It is seen that the growth in defects found is very linear after a branch coverage of about 25%. The testing was terminated at branch coverage of 71% with 20,000 tests applied because no additional defects were found after having applied 1240 tests. The branches not covered were part of the code that would get exercised only rarely (Pasquini et al. 1996). Had the testing continued, the model projects would have found about 43 defects, provided all of the code is reachable.

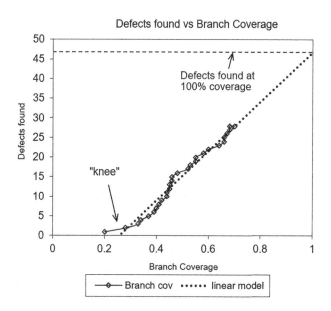

FIGURE 7.3   Coverage-based modeling.

The value of C_{knee}^i is significant. The model in Equations 7.29 and 7.30 suggest that very few defects are detected until the knee is encountered. After the knee, defects found rise linearly with the rise in coverage. In Pasquini's data, the knee occurs at approximately 25% coverage, as can be seen in Figure 7.3. It can be shown (Malaiya and Denton 1998) that the knee occurs at this value

$$C_{knee}^i = 1 - ZD_0 \qquad (7.31)$$

where the parameter Z depends on the attributes of the fault exposure ratio and the corresponding coverage item exposure ratio. Equation 7.31 suggests that the knee occurs very early when the defect density is high. After some testing, the faults that are easier to detect are removed and the remaining faults would only get detected at a higher coverage level. Thus the knee shifts to the higher coverage side when the initial defect density is low.

Defect density and failure rate. Pasquini et al. (1996) have also collected data for the same project that allows computation of the failure rate (per input applied) when each fault found is removed. It is given in Figure 7.4.

From Figure 7.4 it can be noted that code almost always fails until the first seven defects are removed. Removing the first six defects, one after the other, does not significantly decrease the failure rate. It is likely that these faults have a significant test correlation, that is, they are triggered by many common tests, and thus removing one of the first faults makes no significant difference in the failure rate. That should be expected for faults that have a very high testability. This demonstrates that Equation 7.15 for the system failure rate does not apply when the individual failure rates are very high; it would, however, be a good approximation when detectability of the remaining faults is low. At the lower defect density end, the failure rate decreases only gradually as seen at defect densities of about 3. That is because of the fact that these faults have very low testability and thus they are triggered by very few tests.

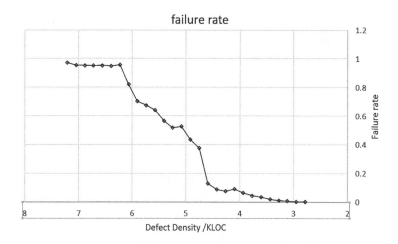

FIGURE 7.4   Failure rate variation with defect density.

Coverage versus mutation testing. In mutation testing, defects (termed mutants) are automatically injected to evaluate the effectiveness of a test strategy. The number of mutants detected then can be taken as a measure of test effectiveness. Its key limitation is that the injected faults may not represent a realistic distribution of faults, especially at low defect densities. On the other hand when branch coverage is measured, covering a branch does not necessarily imply certain detection of all the associated defects. Approaches have been proposed that uses coverage to make mutation testing more efficient (Oliveira et al. 2018), and conversely mutation has been used to make coverage more effective.

7.7 ACHIEVABILITY OF HIGH RELIABILITY IN SOFTWARE

Developing techniques for achieving high reliability in software has long been the aim of the researchers in the field. In real projects, developers face deadlines for getting the software ready for release. The challenge thus is to achieve high reliability within a reasonably short time. The potential approaches can be classified as follows.

1. *Low defect density by design*: Use of tools and development discipline can reduce the initial defect density. The techniques include the following.

   • High-level development: Developing software at a high level and using automatic translation can reduce defect density. Assembly language code is more defect-prone, and with modern optimizing compilers, the need to write time-critical code in assembly has been significantly reduced. Using well-tested library code for common functions (such as graphics) reduces the need to write the corresponding functions in a programming language. Reusing an existing code component from an earlier version is likely to have a lower defect density, provided its functionality and interface are well defined.

   • Integrated development environments (IDEs): IDEs allow better visualization, use of breakpoints for debugging and use of refactoring to automate code modifications. Continuous compilation virtually eliminates syntax errors.

   • Compliance tools: Tools such as those that automatically detect actual or potential memory leaks can reduce some troublesome run-time issues.

2. *Effective testing*: Testing can significantly reduce the number of bugs and the failure rate. Increasing reliability using testing becomes more expensive as the remaining bugs become harder to find. Random testing becomes increasingly ineffective. For very high reliability, the major approaches that can help are (a) use coverage-based testing that uses the structural information and (b) testing for rarely occurring input combinations.

3. *Redundant design*: There have been some investigations into the use of redundancy to implement fault tolerance.

Experiments have found that there is a significant correlation among redundant implementations (N-version programming). Hatton (1997) provided a simple analysis using the experimental data obtained by Knight and Leveson. Knight and Leveson found these probabilities for an input transaction:

- A version failing: 0.0004

- Any two modules failing at the same time (correlated failures): 2.5×10^{-6}

- Three versions failing at the same time (correlated failures): 2.5×10^{-7}

Were all the failures statistically independent, the probability failure of a triple modular redundancy scheme with voting failing would be

$$\text{Failure probability} = \Pr\{\text{all three fail}\} + \Pr\{\text{any two fail}\}$$
$$= (0.0004)^3 + 3(1 - 0.0004)(0.0004)^2$$
$$= 4.8 \times 10^{-7}$$

In the presence of correlation, the probability of failing would be higher:

$$\text{Failure probability} = 2.5 \times 10^{-7} + 3 \times 2.5 \times 10^{-6} = 7.75 \times 10^{-6}$$

Thus while an improvement factor due to redundancy of $0.0004/4.8 \times 10^{-7} = 833.3$ is not achievable, an improvement by $0.0004/7.75 \times 10^{-6} = 51.6$ is still achievable. Hatton argues that none of the testing approaches can reduce the defect density by that factor, and thus redundant design may be an alternative worth considering in some critical situations. It should be noted that Neufelder (2007) found that on average testing reduces the defect density by only a factor of 5.1. Triple modular redundancy increases the cost by a factor of more than 3 (considering the overhead of voting mechanism) but may be considered for systems that need to be highly reliable.

Is ultra-reliable software possible? Butler and Finelli (1993) have argued that it would be hard to quantify the reliability of ultra-reliable software just by testing using the operational profile because the number of failures recorded within a reasonable time would not be statistically significant. They also argue that using probabilistic testing approaches, such as those assumed by common SRGMs, would not be able to achieve a failure rate of 10^{-7} per hour or better.

Probabilistic testing methods lose effectiveness when they are used to further reduce the failure rates or the defect densities to very low values. Some form of directed testing would need to be used involving structural test coverage or fine-grained functional partitioning to apply rarely used input combinations (Hecht 1993).

Is fault-free software possible? There have been claims of fault-free software having been achieved. For example, it has been claimed that "a few projects—for example, the space-shuttle software—have achieved a level of 0 defects in 500,000 lines of code using a system of format development methods, peer reviews, and statistical testing" (McConnell 2004). This is however misleading. The space shuttle software was a project that lasted 30 years from 1981 to 2011, and the last three versions were found to have one error each (Fishman 1996). It has been claimed that formal methods may yield defect-free software. However, a study by Groote et al. (2011) found that the use of formal methods was able to obtain defect densities as low as 0.5 per KLOC. By comparison, the original space shuttle software was found to have a defect density of 0.1 per KLOC, which is regarded as a verified standard (Binder 1997). Formal methods are infeasible for most projects because of the effort and the high degree of expertise required, but may help in special cases.

7.8 CONCLUSION AND FUTURE WORK

We have critically examined the impact of software testing by examining the mathematical modeling approaches using the testability of defects. These include both time-based as well as coverage-based approaches. It is noted that as testing progresses, the remaining faults become harder and harder to find with random testing. For achieving low defect densities more efficiently, structural testing and low-level partitioning need to be used. This may be especially important for finding security vulnerabilities that are security-related defects, some of which can be very hard to find. The mean time a vulnerability remains undetected has recently been found to be 5.7 years (Ablon and Bogart 2017). Finding them sooner is a major challenge. Testing constitutes a major fraction of the development and maintenance costs. Methods for making testing more effective, for accurately modeling their behavior and the tools for automating software testing need to be continually refined to approach the elusive aim of defect-free software.

REFERENCES

Ablon, L., and A. Bogart. *Zero Days, Thousands of Nights: The Life and Times of Zero-Day Vulnerabilities and Their Exploits.* Rand Corporation, 2017.

Adams, E. N. "Optimizing preventive service of software products." *IBM Journal of Research and Development* 28, no. 1, 1984: 2–14.

Binder, R. V. "Can a manufacturing quality model work for software?" *IEEE Software* 14, no. 5, 1997: 101–102.

Boehm, B. W. "A spiral model of software development and enhancement." *Computer* 21, no. 5, 1988: 61–72.

Butler, R. W., and G. B. Finelli. "The infeasibility of quantifying the reliability of life-critical real-time software." *IEEE Transactions on Software Engineering* 19, no. 1, 1993: 3–12.

de Oliveira, A., A. Lobo, C. G. Camilo-Junior, E. N. de Andrade Freitas, and A. M. R. Vincenzi. "FTMES: A failed-test-oriented mutant execution strategy for mutation-based fault localization." *2018 IEEE 29th International Symposium on Software Reliability Engineering (ISSRE)*, pp. 155–165. IEEE, 2018.

Elbaum, S., and S. Narla. "A methodology for operational profile refinement." *2001 Proceedings of the Annual Reliability and Maintainability Symposium*, pp. 142–149. IEEE, 2001.

Fishman, C. "They Write the Right Stuff," *Fast Company*, December 31, 1996. Retrieved February 28, 2019, from https://www.fastcompany. com/28121/they-write-right-stuff.

Groote, J. F., A. Osaiweran, and J. H. Wesselius. "Analyzing the effects of formal methods on the development of industrial control software." *2011 27th IEEE International Conference on Software Maintenance (ICSM)*, Williamsburg, VI, 2011: 467–472.

Le Guen H., and T. Thelin. "Practical experiences with statistical usage testing." *Eleventh Annual International Workshop on Software Technology and Engineering Practice*, Amsterdam, pp. 87–93. 2003.

Hatton, L. "N-version design versus one good version." *IEEE Software* 14, no. 6, 1997: 71–76.

Hecht, H. "Rare conditions: An important cause of failures." *Proceedings of the Eighth Annual Conference on Computer Assurance, 1993*, pp. 81–85. IEEE, 1993.

Horgan, J., and A. Mathur. "Software testing and reliability." In: *The Handbook of Software Reliability Engineering*, pp. 531–565. IEEE Computer Society Press, 1996.

Huang, C.-Y., and C.-T. Lin. "Analysis of software reliability modeling considering testing compression factor and failure-to-fault relationship." *IEEE Transactions on Computers* 59, no. 2, 2010.

Kansal, Y., P. K. Kapur, and U. Kumar. "Coverage-based vulnerability discovery modeling to optimize disclosure time using multiattribute approach." *Quality and Reliability Engineering International* 35, no. 1, 2018: 62–73.

Li, N., and Y. K. Malaiya. "On input profile selection for software testing," *Proceedings of the International Symposium of Software Reliability Engineering*, November 1994, pp. 196–205.

Li, N., and Y. K. Malaiya. "Fault exposure ratio estimation and applications." *Seventh International Symposium on Software Reliability Engineering, 1996. Proceedings*, pp. 372–381. IEEE, 1996.

Malaiya, Y. K. "Reliability allocation." In: *Wiley Encyclopedia of Operations research and Management Science*, John Wiley & Sons, January 14, 2011.

Malaiya, Y. K. "Software Reliability: A Quantitative Approach." In: *System Reliability Management*, pp. 221–252. CRC Press, 2018.

Malaiya, Y. K., and J. Denton. "What do the software reliability growth model parameters represent?" *Proceedings of the Eighth International Symposium on Software Reliability Engineering, 1997*, pp. 124–135. IEEE, 1997.

Malaiya, Y. K., and J. Denton. "Estimating the number of residual defects." *Proceedings of the Third IEEE International High-Assurance Systems Engineering Symposium*, pp. 98–105. IEEE, 1998.

Malaiya, Y. K., N. Karunanithi, and P. Verma. "Predictability of software-reliability models." *IEEE Transactions on Reliability* 41, no. 4, 1992: 539–546.

Malaiya, Y. K., N. Li, J. Bieman, and R. Karcich. "Software test coverage and reliability." *IEEE Transactions on Reliability* December 2002, 420–426.

Malaiya, Y. K., A. Von Mayrhauser, and P. K. Srimani. "An examination of fault exposure ratio." *IEEE Transactions on Software Engineering* 11, 1993: 1087–1094.

Malaiya, Y. K., and S. Yang. "The coverage problem for random testing." *Proceedings of the 1984 International Test Conference on The Three Faces of Test: Design, Characterization, Production*, pp. 237–245. IEEE Computer Society, 1984.

McConnell, S. *Code Complete*. Pearson Education, 2004.

Musa, J. D. "Operational profiles in software-reliability engineering." *IEEE Software* 2, 1993: 14–32.

Musa, J. D., A. Iannino, and K. Okumoto. *Software Reliability: Measurement, Prediction, Application*. McGraw Hill, 1987.

Neufelder, A. M. "Current Defect Density Statistics", 2007, http://www.softrel.com/Current%20defect%20density%20statistics.pdf.

Pasquini, A., A. N. Crespo, and P. Matrella. "Sensitivity of reliability-growth models to operational profile errors vs. testing accuracy [software testing]." *IEEE Transactions on Reliability* 45, no. 4, 1996: 531–540.

Peng, R., Y. F. Li, W. J. Zhang, and Q. P. Hu. "Testing effort dependent software reliability model for imperfect debugging process considering both detection and correction." *Reliability Engineering & System Safety* 126, 2014: 37–43.

Rawat, S., N. Goyal, and M. Ram. "Software reliability growth modeling for agile software development." *International Journal of Applied Mathematics and Computer Science* 27, no. 4, 2017: 777–783.

Regnell, B., P. Runeson, and C. Wohlin, "Towards integration of use case modelling and usage-based testing." *Journal of Software and Systems* 50, no. 2, 2000: 117–130.

Runeson, P., and C. Wohlin, "Statistical usage testing for software reliability control", *Informatica*, 19, No. 2, 1995: 195–207.

Wagner, K. D., C. K. Chin, and E. J. McCluskey. "Pseudorandom testing." *IEEE Transactions on Computers* 3, 1987: 332–343.

Weyuker, E. J. "More experience with data flow testing." *IEEE Transactions on Software Engineering* 19, no. 9, 1993: 912–919.

Yamada, S. *Software Reliability Modeling: Fundamentals and Applications.* Vol. 5. Springer, 2014.

Index

Taylor & Francis Group
an **informa** business

Taylor & Francis eBooks

www.taylorfrancis.com

A single destination for eBooks from Taylor & Francis
with increased functionality and an improved user
experience to meet the needs of our customers.

90,000+ eBooks of award-winning academic content in
Humanities, Social Science, Science, Technology, Engineering,
and Medical written by a global network of editors and authors.

TAYLOR & FRANCIS EBOOKS OFFERS:

A streamlined
experience for
our library
customers

A single point
of discovery
for all of our
eBook content

Improved
search and
discovery of
content at both
book and
chapter level

REQUEST A FREE TRIAL
support@taylorfrancis.com

Routledge
Taylor & Francis Group

CRC CRC Press
Taylor & Francis Group

Printed and bound by CPI Group (UK) Ltd, Croydon, CR0 4YY

01/11/2024

01782621-0002